WORLD GEOGRAPHY

WORLD GEOGRAPHY

Dr. Anoop Kumar Singh

RANDOM PUBLICATIONS

NEW DELHI - 110 002 (INDIA)

World Geography

ISBN 978-93-51117-39-1

Published in 2015 in India by
RANDOM PUBLICATIONS
4376-A/4B, Gali Murari Lal, Ansari Road
New Delhi-110 002
Phone: +9111-43580356, 23289044
E-mail: randomexports@gmail.com; sales@randompublications.com; info@randompublications.com

Reprinted 2024

Type Setting by: Friends Media, Delhi-110089
Digitally Printed at : Replika Press Pvt. Ltd.

Preface

World is a common name for the whole of human civilization, specifically human experience, history, or the human condition in general, worldwide, i.e. anywhere on Earth or pertaining to anywhere on earth. Traditionally, geographers have been viewed the same way as cartographers and people who study place names and numbers. Although many geographers are trained in toponymy and cartology, this is not their main preoccupation. Geographers study the spatial and the temporal distribution of phenomena, processes, and features as well as the interaction of humans and their environment. Because space and place affect a variety of topics, such as economics, health, climate, plants and animals; geography is highly interdisciplinary. The interdisciplinary nature of the geographical approach depends on an attentiveness to the relationship between physical and human phenomena and its spatial patterns.

Geography as a discipline can be split broadly into two main subsidiary fields: human geography and physical geography. The former largely focuses on the built environment and how humans create, view, manage, and influence space. The latter examines the natural environment, and how organisms, climate, soil, water, and landforms produce and interact. The difference between these approaches led to a third field, environmental geography, which combines the physical and the human geography, and looks at the interactions between the environment and humans. Human geography is one of the two major sub-fields of the discipline of geography. Human geography is a branch of the social sciences that studies the world, its people, communities and cultures with an emphasis on relations of and across space and place. Human geography differs from physical geography mainly in that it has a greater focus on studying human activities and is more receptive to qualitative research methodologies. As a discipline, human geography is particularly diverse with respect to its methods and theoretical approaches to study. Geographical knowledge, both physical and social, has a long history. In the history of geography, geographers have often recorded and described features of the Earth that might now be considered the remit of human, rather than physical, geographers. For example Hecataeus of Miletus, a geographer and historian in ancient Greece, described inhabitants of the ancient world as well as physical features. Political geography is the field of human geography that is concerned with the study of both the spatially uneven outcomes of political processes and the ways in which political processes are themselves affected by spatial structures. Conventionally political geography adopts a three-scale structure for the purposes of analysis with the study of the state at the centre, above this is the study of international relations (or geopolitics), and below it is the study of localities. The primary concerns of the sub-discipline can be summarised as the inter-relationships between people, state, and territory. With roots in exploration, map making, and accounts of differences from place to place, modern geography investigates the spatial organization and material character of planet Earth and the physical and human processes that shape its places and landscapes. Geographers study many different topics and places, but they share a concern with what can be learned from analysing spatial arrangements.

This insightful text presents world regional geography as a way to comprehend the world's connections and continuous interplay between features of place, populations, resources, cultures, and political systems.

I thank all members of my team who have helped in the preparation of the book. My special thanks go to "Random Publications" who have published the book.

— Dr. Anoop Kumar Singh

Contents

Chapter 1

World Political and Economic Geography

Introduction

Political geography is the field of human geography that is concerned with the study of both the spatially uneven outcomes of political processes and the ways in which political processes are themselves affected by spatial structures. Conventionally political geography adopts a three-scale structure for the purposes of analysis with the study of the state at the centre, above this is the study of international relations (or geopolitics), and below it is the study of localities. The primary concerns of the sub-discipline can be summarised as the inter-relationships between people, state, and territory.

History

The origins of political geography lie in the origins of human geography itself and the early practitioners were concerned mainly with the military and political consequences of the relationships between physical geography, state territories, and state power. In particular there was a close association with regional geography, with its focus on the unique characteristics of regions, and environmental determinism with its emphasis on the influence of the physical environment on human activities. This association found expression in the work of the German geographer Friedrich Ratzel who, in 1897 in his book *Politische Geographie*, developed the concept of Lebensraum (living space) which explicitly linked the cultural growth of a nation with territorial expansion, and which was later used to provide academic legitimation for the imperialist expansion of the German Third Reich in the 1930s.

The British geographer Halford Mackinder was also heavily influenced by environmental determinism and in developing his concept of the 'geopolitical pivot of history' or heartland (first developed in 1904) he argued that the era of sea power was coming to an end and that land based powers were in the ascendant, and, in particular, that whoever controlled the heartland of 'Euro-Asia' would control the world. This theory involved concepts diametrically opposed to the ideas of Alfred Thayer Mahan about the significance of *sea power* in world conflict. The heartland theory hypothesized the possibility of a huge empire being created which didn't need to use coastal or transoceanic transport to supply its military industrial complex, and that this empire could not be defeated by the rest of the world coalitioned against it. This perspective proved influential throughout the period of the Cold War, underpinning military thinking about the creation of buffer states between East and West in central Europe.

The heartland theory depicted a world divided into a *Heartland* (Eastern Europe/Western Russia); *World Island* (Eurasia and Africa); *Peripheral Islands* (British Isles, Japan, Indonesia and Australia) and *New World* (The Americas). Mackinder claimed that whoever controlled the Heartland would have control of the world. He used this warning to politically influence events such as the Treaty of Versailles, where buffer states were created between the USSR and Germany, to prevent either of the them controlling the Heartland. At the same time, Ratzel was creating a theory of states based around the concepts of Lebensraum and Social Darwinism. He argued that states were 'organisms' that needed sufficient room in which to live. Both of these writers created the idea of a political and geographical science, with an objective view of the world. Pre-World War II political geography was concerned largely with these issues of global power struggles and influencing state policy, and the above theories were taken on board by German geopoliticians such as Karl Haushofer who-perhaps inadvertently-greatly influenced Nazi political theory. A form of politics legitimated by 'scientific' theories such as a 'neutral' requirement for state expansion was very influential at this time.

The close association with environmental determinism and the freezing of political boundaries during the Cold War led to a considerable decline in the importance of political geography which was described by Brian Berry in 1968 as 'a moribund backwater'. Although in other areas of human geography a number of new approaches were invigorating research, including quantitative spatial science, behavioural studies, and structural marxism, these were largely ignored by political geographers whose main point of reference continued to be the regional approach. As a result much political geography of this period was descriptive with little attempt to produce generalisations from the data collected. It was not until 1976 that Richard Muir could argue that political geography might not be a dead duck but could in fact be a phoenix.

Areas of Study

From the late-1970s onwards political geography has undergone a renaissance, and could fairly be described as one of the most dynamic of the sub-disciplines today. The revival was underpinned by the launch of the journal Political Geography Quarterly (and its expansion to bi-monthly production as Political Geography). In part this growth has been associated with the adoption by political geographers of the approaches taken up earlier in other areas of human geography, for example, Ron J. Johnston's (1979) work on electoral geography relied heavily on the adoption of quantitative spatial science, Robert Sack's (1986) work on territoriality was based on the behavioural approach, and Peter Taylor's (e.g. 2007) work on World Systems Theory owes much to developments within structural Marxism. However the recent growth in the vitality and importance of the sub-discipline is also related to changes in the world as a result of the end of the Cold War, including the emergence of a new world order (which as yet is only poorly defined), and the development of new research agendas, such as the more recent focus on social movements and political struggles going beyond the study of nationalism with its explicit territorial basis. Recently, too, there has been increasing interest in the geography of green politics, including the geopolitics of environmental protest, and in the capacity of our existing state apparatus and wider political institutions to address contemporary and future environmental problems competently.

Political geography has extended the scope of traditional political science approaches by acknowledging that the exercise of power is not restricted to states and bureaucracies, but is part of everyday life.

This has resulted in the concerns of political geography increasingly overlapping with those of other human geography sub-disciplines such as economic geography, and, particularly, with those of social and cultural geography in relation to the study of the politics of place. Although contemporary political geography maintains many of its traditional concerns the multi-disciplinary expansion into related areas is part of a general process within human geography which involves the blurring of boundaries between formerly discrete areas of study, and through which the discipline as a whole is enriched.

In particular, then, modern political geography often considers:

- How and why states are organized into regional groupings, both formally (e.g. the European Union) and informally (e.g. the Third World)
- The relationship between states and former colonies, and how these are propagated over time, for example through neo-colonialism
- The relationship between a government and its people
- The relationships between states including international trades and treaties
- The functions, demarcations and policings of boundaries
- How imagined geographies have political implications
- The influence of political power on geographical space
- The study of election results (electoral geography).

Critical Political Geography

Critical political geography is mainly concerned with the criticism of traditional political geographies. As with much of the move towards 'Critical geographies', the arguments have drawn largely from postmodern, poststructural and postcolonial theories. Examples include:

- Feminist geography, which argues for a recognition of the power relations as patriarchal and attempts to theorise alternative conceptions of identity and identity politics. Alongside related concerns such as Queer theory and Youth studies
- Postcolonial theories which recognise the Imperialistic, universalising nature of much political geography, especially in Development geography.

Ceylon: Some Problems of a Plural Society

The Sri Lankan Civil War was a conflict fought on the island of Sri Lanka. Beginning on July 23, 1983, there was an on-and-off insurgency against the government by the Liberation Tigers of Tamil Eelam (the LTTE, also known as the Tamil Tigers), a separatist militant organization which fought to create an independent Tamil state named Tamil Eelam in the north and the east of the island. After a 30 year long military campaign, the Sri Lankan military defeated the Tamil Tigers in May 2009.

For over 25 years, the insurgency caused significant hardships for the population, environment and the economy of the country, with an estimated 80,000-100,000 people killed during its course. The tactics employed by the Liberation Tigers of Tamil Eelam resulted in their being branded as a terrorist organization in 32 countries, including the United States, India, Australia, Canada and the member nations of the European Union.

After two decades of fighting and three failed attempts at peace talks, including the unsuccessful deployment of the Indian Army, the Indian Peace Keeping Force from 1987 to 1990, a lasting negotiated settlement to the conflict appeared possible when a cease-fire was declared in December 2001, and a ceasefire agreement signed with international mediation in 2002. However, limited hostilities renewed in late 2005 and the conflict began to escalate until the government launched a number of major military offensives against the LTTE beginning in July 2006, driving the LTTE out of the entire Eastern province of the island. The LTTE then declared they would "resume their freedom struggle to achieve statehood".

In 2007, the government shifted its offensive to the north of the country, and formally announced its withdrawal from the ceasefire agreement on January 2, 2008, alleging that the LTTE violated the agreement over 10,000 times. Since then, aided by the destruction of a number of large arms smuggling vessels that belonged to the LTTE, and an international crackdown on the funding for the Tamil Tigers, the government took control of the entire area previously controlled by the Tamil Tigers, including their de-facto capital Kilinochchi, main military base Mullaitivu and the entire A9 highway, leading the LTTE to finally admit defeat on May 17, 2009. Following the end of the war, the Sri Lankan government has described Sri Lanka as the first country in the modern world to eradicate terrorism on its own soil.

Origin and Evolution

The root of modern conflict goes back to British colonial rule when the country was known as Ceylon. A nationalist political movement from Sinhalese communities arose in the country in the early 20th century with the aim of obtaining political independence, which was eventually granted by the British after peaceful negotiations in 1948. Disagreements between the Sinhalese and Tamil ethnic communities flared up when drawing up the country's first post-independence constitution.

After their election to the State Council in 1936, the Lanka Sama Samaja Party (LSSP) members N. M. Perera and Philip Gunawardena demanded the replacement of English as the official language by Sinhala and Tamil. In November 1936, a motion that 'in the Municipal and Police Courts of the Island the proceedings should be in the vernacular' and that 'entries in police stations should be recorded in the language in which they are originally stated' were passed by the State Council and referred to the Legal Secretary. However, in 1944, J. R. Jayawardene moved in the State Council that Sinhala should replace English as the official language.

In 1956 Prime Minister S. W. R. D. Bandaranaike's passage of the "Sinhala Only Act" led to ethnic riots. The civil war is a direct result of the escalation of the confrontational politics that followed. The formation of the Tamil United Liberation Front (TULF) with its Vaddukkodei (Vattukottai)) resolution of 1976 led to a hardening of attitudes.

In 1963, shortly after the nationalisation of oil companies by the Sri Lankan government, documents relating to a separate Tamil state of 'Tamil Eelam' began to circulate. At this time, Anton Balasingham, an employee of the British High Commission in Colombo, began to participate in separatist activities. He later migrated to Britain, where he became the chief theoretician of the Liberation Tigers of Tamil Eelam. In the late 1960s, several Tamil youth, among them Velupillai Prabhakaran also became involved in these activities. These forces together formed the Tamil New Tigers in 1972. This was formed around an ideology which looked back to the 1st Millennium Chola Empire-the Tiger was the emblem of that empire.

A further movement, the Eelam Revolutionary Organisation of Students, formed in Manchester and London; it became the backbone of the Eelamist movement in the diaspora, arranging passports and employment for immigrants and levying a heavy tax on them. It became the basis of the Eelamist logistical organisation, later taken over entirely by the Liberation Tigers of Tamil Eelam.

The TULF supported the armed actions of the young militants of the TNT who were dubbed "our boys. " These "boys" were the product of the post-war population explosion. Many partially educated, unemployed Tamil youth fell for revolutionary solutions to their problems. The leftist parties had remained "non-communal" for a long time, but the Federal Party (as well as its off-shoot, the TULF), deeply conservative and dominated by Vellala casteism, did not attempt to form a national alliance with the leftists in their fight for language rights.

Following the sweeping electoral victory of the UNP in July 1977, the TULF became the leading opposition party, with around one sixth of the total electoral vote winning on a party platform of secession from Sri Lanka.

In August 1977, Junius Richard Jayawardene's new UNP government followed its attack on the Left with a well organised pogrom against Tamils living in majority Sinhalese areas. In August the government granted only the educational rights demanded by the Tamils. But to the Tamil leadership that was losing the control it had on the Tamil militants after not being able to follow through with the election promise of seceding from Sri Lanka to form Tamil, it was too little too late.

Outbreak of Civil War

Supported by the on-going politics of conflict in Sri Lanka, politicized Tamil youth in the North and the East started to form militant groups. These groups developed independently of the Colombo Tamil leadership, and in the end rejected and annihilated them. The most prominent of these groups was the TNT, which changed its name to the Liberation Tigers of Tamil Eelam or the LTTE in 1976. The LTTE initially carried out a campaign of violence against the state, particularly targeting policemen and also moderate Tamil politicians who attempted a dialogue with the government. Their first major operation was the assassination of the mayor of Jaffna, Alfred Duraiappah in 1975 by Prabhakaran.

The LTTE 's *modus operandi* of the early war was based on assassinations. The assassination in 1977 of a Tamil Member of Parliament, M. Canagaratnam, was carried out personally by Prabhakaran, the leader of the LTTE.

In July 1983, the LTTE launched a deadly attack on the military in the North of the country, killing 13 soldiers. Using the nationalistic sentiments to their advantage, the Jayawardena now organized massacres and pogroms in Colombo, the capital, and elsewhere-although several pogroms had been reported in rural areas prior to the LTTE attack. Between 400 and 3,000 Tamils were estimated to have been killed, and many more fled Sinhalese-majority areas. This is usually considered the beginning of the civil war.

Apart from the LTTE, there initially was a plethora of militant groups. The LTTE's position, adopted from that of the PLO, was that there should be only one. Initially the LTTE gained prominence due to devastating attacks such as the massacre of civilians at the Kent and Dollar Farms in 1984 and the Anuradhapura massacre of 146 civilians in 1985. The Anuradhapura massacre was apparently retaliated for by government forces with the Kumudini boat massacre in which over 23 Tamil civilians died. Over time the LTTE merged with or largely exterminated almost all the other militant Tamil groups.

As a result, many Tamil splinter groups ended up working with the Sri Lankan government as paramilitaries or denounced violence and joined mainstream politics, and some legitimate Tamil-oriented political parties remain, all opposed to LTTE's vision of an independent state.

Peace talks between the LTTE and the government began in Thimphu in 1985, but they soon failed, and the war continued. In 1986 many civilians were massacred as part of this conflict. In 1987, government troops pushed the LTTE fighters to the northern city of Jaffna. In April 1987, the conflict exploded with ferocity, as both the government forces and the LTTE fighters engaged each other in a series of bloody operations.

The Sri Lankan military launched an offensive, called "Operation Liberation" or "Vadamarachchi Operation", during May–June 1987, to regain the territory in Jaffna peninsula from the LTTE's control. This offensive marked the Sri Lankan military's first ever conventional warfare in Sri Lankan soil since independence.

The military offensive was successful and the LTTE leader Prabhakaran and the Sea Tiger leader Soosai narrowly escaped from advancing troops at Valvettithurai. The key military personnel involved in the operation were Lt. Col. Vipul Boteju, Lt Col. Sarath Jayawardane, Col. Vijaya Wimalaratne, Brig. Denzil Kobbekaduwa and Maj Gotabaya Rajapaksa.

In July 1987, the LTTE carried out their first suicide attack: "Captain Miller" of the Black Tigers drove a small truck with explosives through the wall of a fortified Sri Lankan army camp, reportedly killing forty soldiers. Since then they have carried out over 170 suicide attacks, more than any other organization in the world, and the suicide attack has become a trademark of the LTTE, and a characteristic of the civil war.

The killings of Father Mary Bastian and George Jeyarajasingham, both human rights activists, have been attributed to the government forces. These deaths are examples of thousands murdered that happened in this period.

Indian Involvement

India became involved in the conflict in the 1980s for a number of reasons, including its leaders' desire to project India as the regional power in the area and worries about India's own Tamils seeking independence. The latter was particularly strong in the Indian state of Tamil Nadu, where ethnic kinship led to strong support for independence for Sri Lankan Tamils. Throughout the conflict, the Indian central and state governments have supported both sides in different ways. Beginning in the 1980s, India, through its intelligence agency R&AW, provided arms, training and monetary support to a number of Sri Lankan Tamil militant groups, including the LTTE and its rival Tamil Eelam Liberation Organization (TELO). The LTTE's rise is widely attributed to the initial backing it received from R&AW. It is believed that by supporting different militant groups, the Indian government hoped to keep the Tamil independence movement divided and be able to exert overt control over it.

India became more actively involved in the late 1980s, and on June 5, 1987, the Indian Air Force airdropped food parcels to Jaffna while it was under siege by Sri Lankan forces. At a time when the Sri Lankan government stated they were close to defeating the LTTE, India dropped 25 tons of food and medicine by parachute into areas held by the LTTE in a direct move of support toward the rebels. Negotiations were held, and the Indo-Sri Lanka Peace Accord was signed on July 29, 1987, by Indian Prime Minister Rajiv Gandhi and Sri Lankan President Jayewardene. Under this accord, the Sri Lankan Government made a number of concessions to Tamil demands, including a devolution of power to the provinces, a merger—subject to later referendum—of the Northern and the Eastern provinces into the single province, and official status for the Tamil language (this was enacted as the 13th Amendment to the Constitution of Sri Lanka). India agreed to establish order in the North and East through a force dubbed the Indian Peace Keeping Force (IPKF), and to cease assisting Tamil insurgents. Militant groups including the LTTE, although initially reluctant, agreed to surrender their arms to the IPKF, which initially oversaw a cease-fire and a modest disarmament of the militant groups.

The signing of the Indo-Lanka Accord, so soon after JR Jayawardene's declaration that he would fight the Indians to the last bullet, led to unrest in south. The arrival of the IPKF to take over control of most areas in the North of the country enabled the Sri Lanka government to shift its forces to the south (in Indian aircraft) to quell the protests. This led to an uprising by the Janatha Vimukthi Peramuna in the south, which was put down bloodily over the next two years.

While most Tamil militant groups laid down their weapons and agreed to seek a peaceful solution to the conflict, the LTTE refused to disarm its fighters. Keen to ensure the success of the accord, the IPKF then tried to demobilize the LTTE by force and ended up in full-scale conflict with them. The three year long conflict was also marked by the IPKF being accused of committing various abuses of human rights by many human rights groups as well as some within the Indian media.

The IPKF also soon met stiff opposition from the Tamils. Simultaneously, nationalist sentiment led many Sinhalese to oppose the continued Indian presence in Sri Lanka. These led to the Sri Lankan government's call for India to quit the island, and they allegedly entered into a secret deal with the LTTE that culminated in a ceasefire. The LTTE and IPKF continued to have frequent hostilities, and according to some reports, the Sri Lankan government even armed the rebels in order to see the back of the Indian forces. Although casualties among the IPKF mounted, and calls for the withdrawal of the IPKF from both sides of the Sri Lankan conflict grew, Gandhi refused to remove the IPKF from Sri Lanka. However, following his defeat in Indian parliamentary elections in December 1989, the new prime Minister V. P. Singh ordered the withdrawal of the IPKF, and their last ship left Sri Lanka on March 24, 1990. The 32 month presence of the IPKF in Sri Lanka resulted in the deaths of 1100 Indian soldiers and over 5000 Sri Lankans. The cost for the Indian government was estimated at over 20 billion rupees.

Rajiv Gandhi's Assassination

Support for the LTTE in India dropped considerably in 1991, after the assassination of ex-Prime Minister Rajiv Gandhi by a suicide bomber Thenmuli Rajaratnam. The Indian press has subsequently reported that Prabhakaran decided to eliminate Gandhi as he considered Gandhi to be against the Tamil liberation struggle and feared that Gandhi might re-induct the IPKF, which Prabhakaran termed the "satanic force", if he won the 1991 Indian elections. In 1998 a court in India presided over by Special Judge V. Navaneetham found the LTTE and its leader Velupillai Prabhakaran responsible for the assassination and in a 2006 interview, LTTE ideologue Anton Balasingham stated regret over the assassination, although he stopped short of outright acceptance of responsibility for it.

India remains an outside observer to the ongoing peace process, with frequent demands by many groups for an extradition of Velupillai Prabhakaran, now deceased.

Pan-Tamil Support

In 2008, the central coalition in India was rocked by threats, resignations, and arrests based on Tamil nationalism, and hence support to Tamils in Sri Lanka. Following the historically pro-Tamil DMK party's accession to power in Tamil Nadu and the centre, it was seen as though there would be more political support from India. In 2008, party chief and TN CM Karunanidhi accepted the resignation of multiple MP's of his party in protest against an increasing casualty count of Tamil civilians in the war. Following this, MDMK founder and general secretary, Vaiko, courted arrest on charges of sedition in saying he would take up arms to fight on the side of the Tamils. He then charged the Indian Government with abetting the Sri Lankan Government in order to eliminate the Tamils there. He added that Sri Lanka would heed a request for a ceasefire if India imposed economic sanctions on the country.

In a rare show of unanimity, all the parties in Tamil Nadu assembly unanimously demanded a ceasefire in conflict, while appealing to the Centre to make efforts to stop the Sri Lankan military offensive.

Even the Congress party, which had seen the issue as an untouchable subject for more than a decade, said there could be no two opinions on the need for a ceasefire. To this, party floor leader, D Sudarsanam, said that the Centre was making efforts to stop the war and the results would soon be known. Congress whip, Peter Alphonse, denied that his party was acting against the interests of the Sri Lankan Tamils and said he was ready to list his party's efforts for the welfare of the said Tamils. The deputy leader of the opposition and senior AIADMK leader, O Panneerselvam, made a charge that the "intransigent attitude" of the Sri Lankan government was the reason for the continuation of the war. He added that the Sri Lankan army was bombing schools and public places that had resulted in the deaths of innocent people, including children.

The LTTE, which was cornered in a tiny section of the island, desperately pleaded with the Indian Tamil parties to demand a ceasefire, in order to save their leadership from annihilation, which led to several Indian Tamil politician requesting the Sri Lankan Government to call for an immediate ceasefire.

Eelam War II

In the 1980s and 1990s, successive governments enacted a number of official acts to appease the Tamil community, including recognizing Tamil as an official language and merging the Northern and the Eastern Provinces of the country.

Yet the violence continued, as the LTTE continued to massacre innocent villagers living in the deep rural Eastern and North-Eastern areas. Dollar and Kent Farm massacres, where hundreds of men, women and children were attacked during the night as they slept and were hacked to death with fatal blows to the head from axes, are two clear examples of the ruthless acts of terrorism by the LTTE. Some of the men who defended themselves were executed their hands tied behind their backs. The LTTE used these terror tactics to scare Sinhalese and Muslim farmers away from these areas and swiftly took control of significant parts of the North, in a clear act of ethnic cleansing in Sri Lanka. When the Indian Peace Keeping Force withdrew, the LTTE established many government-like functions in the areas under its control. A tentative ceasefire held in 1990 as the LTTE occupied itself with destroying rival Tamil groups while the government cracked down on the JVP uprising. When both major combatants had established their power bases, they turned on each other and the ceasefire broke down. The government launched an offensive to try to retake Jaffna. This phase of the war soon acquired the name *Eelam War II.*

It was marked by unprecedented brutality. The LTTE massacred 113 Sinhalese and Muslim policemen after they had surrendered on promises of safe conduct. The government placed an embargo on food and medicine entering the Jaffna peninsula and the air force relentlessly bombed LTTE targets in the area. The LTTE responded by attacking Sinhalese and Muslim villages and massacring civilians. One of the largest civilian massacres of the war occurred when the LTTE massacred 166 Muslim civilians at Palliyagodella. The government trained and armed Home Guard Muslim units then took revenge on Tamil villages. There was also significant massacre of Tamil civilians attributed to government forces, especially in the Eastern Province. Notable international jurist Neelan Thiruchelvam, in a speech at the ICES-Colombo, indicated that the appropriate investigations into massacres and disappearances of civilians including many children in the Sathurukondan, Eastern University, Mylanthanai and the mass murder and burial of school children at Sooriyakanda were hampered by the adoption of emergency regulations which were contributing to a climate of impunity. Along roadsides in the North and East, burning bodies became a common sight. Throughout the country, government death squads hunted down, kidnapped, or killed Sinhalese or Tamil youth suspected of being JVP or LTTE sympathizers, respectively. In October 1990, the LTTE expelled all the Muslims residing in Jaffna. A total of 28,000 Muslims were forced to leave their homes taking nothing but the clothes on their backs.

The largest battle of the war was in July 1991, when the army's Elephant Pass (Alimankada) base, which controlled access to the Yapanaya (Jaffna) peninsula, was surrounded by 5,000 LTTE troops. More than 2,000 died on both sides in the month-long siege, before 10,000 government troops arrived to relieve the base.

In February 1992, another series of government offensives failed to capture Jaffna due to Indian intervention, which led to a military stand-off as Indian threatened to invade Sri Lanka. The Sri Lankan Government backed off and ordered the withdrawal of its forces, only days away from capturing LTTE supremo Velupillai Prabharan, whilst he lay surrounded in his home town of Vellvetithurai. Lt. General Denzil Kobbekaduwa together with Major General Vijaya Wimalaratne and Rear Admiral Mohan Jayamaha, died on August 8, 1992, at Araly (Aeraella) point Jaffna due to a land mine blast, which badly affected military morale.

The LTTE, for its part, scored a major victory when one of their suicide bombers killed Sri Lankan President Ranasinghe Premadasa in May 1993. In November 1993 the LTTE succeeded in the Battle of Pooneryn.

Eelam War III

In the 1994 parliamentary elections, the UNP was defeated and, amidst great hope, the People's Alliance, headed by Chandrika Kumaratunga, came to power on a peace platform. Chandrika Kumaratunga won the presidential elections as well after the LTTE assassinated the opposition leader Gamini Dissanayake. A ceasefire was agreed in January 1995, but the ensuing negotiations proved fruitless. The LTTE broke the ceasefire on April 19 and thus began the next phase of the war, dubbed *Eelam War III.*

The new government then pursued a policy of "war for peace". Determined to retake the key rebel stronghold of Jaffna, which was occupied by 2,000 rebels, it poured troops into the peninsula. In one particular incident in August 1995, Air Force jets bombed St. Peter's church at Navali (Naavaella), killing at least 65 refugees and wounding 150 others. Government troops initially cut off the peninsula from the rest of the island, and then after 7 weeks of heavy fighting succeeded in bringing Jaffna under government control for the first time in nearly a decade. In a high profile ceremony, Sri Lankan Defense Minister Anurudda Ratwatte raised the national flag inside the Jaffna fort on December 5, 1995. The government estimated that approximately 2500 soldiers and rebels were killed in the offensive, and an estimated 7,000 wounded. Many civilians were killed as part of this conflict such as the Navaly church bombing in which over 125 civilians died. The LTTE and more than 350,000 civilians, compelled by LTTE pressure to leave Jaffna, fled to the Vanni region in the interior. Most of the refugees returned later the next year.

The LTTE responded by launching Operation Unceasing Waves and decisively won the Battle of Mullaitivu in July 1996.

The government launched another offensive in August 1996. Another 200,000 civilians fled the violence. The town of Kilinochchi (Giraa Nikke) was taken on September 29. On May 13, 1997, 20,000 government troops tried to open a supply line through the LTTE-controlled Vanni, but failed. Civilians were regularly killed and wounded by both sides.

As violence continued in the North, LTTE suicide and time bombs were exploded numerous times in populated city areas and public transport in the south of the country, killing hundreds of civilians. In January 1996, the LTTE carried out one of their deadliest suicide bomb attacks at the Central Bank in Colombo, killing 90 and injuring 1,400. In October 1997 they bombed the Sri Lankan World Trade Centre and, in January 1998, detonated a truck bomb in Kandy (Mahanuvara), damaging the Temple of the Tooth, one of the holiest Buddhist shrines in the world. In response to this bombing, the Sri Lankan government outlawed the LTTE and with some success pressed other governments around the world to do the same, significantly interfering with their fund-raising activities.

On September 27, 1998 the LTTE launched the Operation Unceasing Waves II and after heavy fighting captured Kilinochchi, thus winning Battle of Kilinochchi.

In March 1999, in Operation Rana Gosa, the government tried invading the Vanni from the south. The army made some gains, taking control of Oddusuddan (Oththan-thuduva) and Madhu, but could not dislodge the LTTE from the region. In September 1999 the LTTE massacred 50 Sinhalese civilians at Gonagala

The LTTE returned to the offensive with the Operation Unceasing Waves III on November 2, 1999. Nearly all the Vanni rapidly fell back into LTTE hands. The LTTE launched 17 successful attacks in the region which culminated in the overrunning of the Paranthan (Puranthaenna) Chemicals Factory base and the Kurrakkan Kaddukulam (kurakkan-kaela vaeva) base. Thousands were killed in the fighting. The rebels also advanced north towards Elephant Pass (Alimankada) and Jaffna (Yapanaya). The LTTE was successful in cutting all land and sea supply lines of the Sri Lankan armed forces to the south, west and north of the town of Kilinochchi. In December 1999 the LTTE attempted to assassinate President Chandrika Kumaratunga in a suicide attack at a pre-election rally. She lost one eye, among other injuries, but was able to defeat opposition leader

Ranil Wickremesinghe in the Presidential election and was re-elected for her second term in office.

On April 22, 2000, the Elephant Pass military complex, which had separated the Jaffna peninsula from the Vanni mainland for 17 years, completely fell to the hands of the LTTE. The army then launched Operation Agni Khiela to take back the southern Jaffna Peninsula, but sustained losses. The LTTE continued to press towards Jaffna, and many feared it would fall to the LTTE, but the military repulsed LTTE offensives and was able to maintain control of the city.

Early Peace Efforts

Exhaustion with the war was building as casualties mounted and there appeared to be no end in sight. By mid-2000, human rights groups estimated that more than one million people in Sri Lanka were internally displaced persons, living in camps, homeless and struggling for survival. As a result, a significant peace movement developed in the late 1990s, with many organizations holding peace camps, conferences, trainings and peace meditations, and many other efforts to bridge the two sides at all levels. As early as February 2000, Norway was asked to mediate by both sides, and initial international diplomatic moves began to find a negotiated settlement to the conflict.

Hopes for peace gained ground as the LTTE declared a unilateral ceasefire in December 2000, but they cancelled it on April 24, 2001, and launched another offensive against the government. After securing a vast area controlled by the military, the LTTE further advanced northwards. This advancement of the LTTE was posing a serious threat to the Elephant Pass (Alimankada) military complex that housed 17,000 troops of the Sri Lankan forces.

In July 2001 the LTTE carried out a devastating suicide attack on Bandaranaike International Airport, destroying eight of the air force's planes (2 IAI Kfirs, 1 Mil-17, 1 Mil-24, 3 K-8 trainers, 1 MiG-27) and four Sri Lankan Airlines planes (2 Airbus A330s, 1 A340 and 1 A320), dampening the economy and causing tourism, a vital foreign exchange earner for the government, to plummet.

2002 Peace Process

Beginning of the Ceasefire

Towards the end of 2001, however, following the attacks of 9/11, the LTTE began to declare their willingness to explore measures for a peaceful settlement to the conflict. The LTTE are believed to have taken this action after fear of international pressure and even direct US support of the Sri Lankan Government as part of the War on Terror. In the south, the government was facing increasing criticism over its "war for peace" strategy, with peace nowhere in sight, and the economy in tatters. After losing a no-confidence motion, President Kumaratunga was forced to dissolve parliament and call for fresh elections. The elections, held on December 5, 2001, saw a sweeping victory for the United National Front, led by Ranil Wickremasinghe, who campaigned on a pro-peace platform and pledged to find a negotiated settlement to the conflict.

On December 19, amidst efforts by Norway to bring the government and the Tamil Tigers to the negotiating table, the LTTE announced a 30 day ceasefire with the Sri Lankan government and pledged to halt all attacks against government forces. The new government welcomed the move, and reciprocated it 2 days later, announcing a month long ceasefire and agreeing to lift a long standing economic embargo on rebel-held territory.

Signing of Memorandum of Understanding

The two sides formalized a Memorandum of Understanding (MoU) on February 22, 2002, and signed a permanent ceasefire agreement (CFA). Norway was named mediator, and it was decided that they, together with the other Nordic countries, monitor the ceasefire through a committee of experts named the Sri Lanka Monitoring Mission. In August, the government agreed to lift the ban on the LTTE and paved the way for the resumption of direct negotiations with the LTTE.

Following the signing of the ceasefire agreement, commercial air flights to Jaffna began and the LTTE opened the key A9 highway, which linked government controlled area in the south with Jaffna and ran through LTTE territory, allowing civilian traffic through the Vanni region for the first time in many years, but only after paying a tax to the LTTE. Many foreign countries also offered substantial financial support if peace was achieved and optimism grew that an end to the decades-long conflict was in sight.

The much-anticipated peace talks began in Phuket, Thailand on September 16, and 5 further rounds followed in Phuket, Norway and Berlin, Germany. During the talks, both sides agreed to the principle of a federal solution and the Tigers dropped their long-standing demand for separate state. This was a key compromise from the LTTE, which had always insisted on an independent Tamil state and it also represented a compromise from the government, which had seldom agreed to more than minimal devolution. Both sides also exchanged prisoners of war for first time.

Political Changes in the South

Following the elections of 2001, for the first time in Sri Lanka's history, the President and Prime Minister were of two different parties. This co-habitation was uneasy, especially since Prime Minister Wickremasinghe and the UNP favoured a federal solution to the conflict, while hard-line elements within President Kumaratunga's party and other Sinhala nationalist groups allied to her opposed one as they did not trust the LTTE, which continued to levy taxes, strengthen themselves by smuggling in arms and ammunition, recruit child soldiers, and engage in killings of members of rival Tamil groups and government intelligence agents. During this time the LTTE also succeeded in setting up a series of vital bases around the Trincomalee (Gokanna) harbour and the Eastern Province.

The talks broke down on April 21, 2003, when the Tamil Tigers announced they were suspending any further talks due to their "displeasure" at the handling of some "critical issues". Among the reasons the Tigers gave were their exclusion from reconstruction talks in Washington DC on April 14 and a more general insinuation that they were not receiving the full economic rewards of peace. They cited the failure, as they saw it, of peace-dividends to transfer to security withdrawals on the ground and the disparity, as they saw it, between the relative calm of the government-held northeast and continuing violence in Tiger-held areas. However, the LTTE maintained it was committed to a settlement to the two-decade conflict, but stated that progress had to be made on the ground before the settlement proceeded.

On October 31, the LTTE issued its own peace proposal, calling for an *Interim Self Governing Authority* (ISGA). The ISGA would be fully controlled by the LTTE and would have broad powers in the North and East. This provoked a strong backlash among the hardline elements in the South, who accused Prime Minister Wickremasinghe of handing the North and East to the LTTE. Under pressure from within her own party to take action, Kumaratunga declared a state of emergency and took three key government ministries, the Ministry of Mass Media, the Interior Ministry and the crucial Defense Ministry. She then formed an alliance with the JVP, called the United People's Freedom Alliance, opposed to the ISGA and advocating a harder line on the LTTE, and called for fresh elections. The elections, held on April 8, 2004, resulted in victory for the UPFA with Mahinda Rajapakse appointed as Prime Minister. Initial fears of a resumption of the conflict were proved unfounded when the new government expressed its desire to continue the peace process and find a negotiated settlement to the conflict.

Split of the LTTE

Meanwhile, in March 2004 there had been a major fracturing between the northern and eastern wings of the LTTE. Colonel Karuna, the Eastern commander of the LTTE and one of Prabhakaran's trusted lieutenants, pulled 5,000 eastern cadres out of the LTTE, claiming insufficient resources and power were being given to Tamils of the eastern part of the island. It was the biggest expression of dissension in the history of the LTTE and a civil war within the LTTE seemed imminent. After the parliamentary elections, brief fighting south of Trincomalee (formerly known as Gokanna) led to a rapid retreat and capitulation of Karuna's group, their leaders eventually going into hiding including Karuna himself, who was helped to escape by Seyed Ali Zahir Moulana, a powerful politician from the ruling party. However, the "Karuna faction" maintained a significant presence in the East and continued to launch attacks against the LTTE. The LTTE accuses the army of covertly backing the breakaway group, which subsequently formed a political party named the Tamil Eela Makkal Viduthalai Pulikal (TMVP) and hopes to contest in future elections.

The ceasefire largely held through all this turmoil, with over 3000 infractions by the LTTE and some 300 by the SLA recorded by the Sri Lanka Monitoring Mission (SLMM) by 2005. The situation was further complicated by allegations that both sides were carrying out covert operations against each other. The government claimed that the LTTE was killing political opponents, recruiting children, importing arms, and killing government security and intelligence officers. The rebels accused the government of supporting paramilitary groups against them, especially the Karuna group.

Tsunami and Aftermath

On December 26, 2004, the Indian Ocean tsunami hit Sri Lanka, killing more than 30,000 people, and leaving many more homeless. Aid poured in from donor countries, but disagreements arose instantly over how it should be distributed to the Tamil regions under LTTE control. By June 24, the government and LTTE agreed on the Post-Tsunami Operational Management Structure (P-TOMS), but it received sharp criticism from the JVP, who left the government in protest. The legality of P-TOMS was also challenged in the courts. President Kumaratunga eventually had to scrap P-TOMS, which led to widespread criticism that sufficient aid was not reaching the North and East of the country. However, immediately following the tsunami there was a marked decrease in violence in the North.

Sri Lankan Foreign Minister Lakshman Kadirgamar, a Tamil who was highly respected by foreign diplomats and who had been sharply critical of the LTTE, was assassinated at his home on August 12, 2005, allegedly by an LTTE sniper. His assassination led to the marginalization of the LTTE from the international community, and is thought to be the instant when the LTTE lost much of its sympathy in the eyes of foreign nations. Hence the silence of the international community when the Sri Lankan government took military action against the LTTE in 2006, when the latter closed the Mavil Oya (Mavil aru) sluice.

Further political change occurred when the Supreme Court of Sri Lanka declared President Kumaratunga's second and final term over and ordered her to hold fresh presidential elections. The main candidates for the election, which was held in November, were the UNF candidate, former Prime Minister Ranil Wickremasinghe, who advocated the reopening of talks with the LTTE, and the UPFA candidate, Prime Minister Rajapaksa, who called for a tougher line against the LTTE and a renegotiation of the ceasefire. The LTTE openly called for a boycott of the election by the Tamils. Many of them were expected to vote for Wickremasinghe, and the loss of their votes proved fatal to his chances as Rajapakse achieved a narrow win. Following the election, the LTTE leader Velupillai Prabhakaran stated in his annual address that the Tigers would "renew their struggle" in 2006 if the government did not take serious moves toward peace.

Resumption of Hostilities

Just days after Prabhakaran's speech, a new round of violence erupted. Beginning in December 2005, there was increased guerrilla activity to the northeast, including Claymore mine attacks which killed 150 government troops, clashes between the Sea Tigers and the Sri Lanka navy, and the killings of sympathizers on both sides including Taraki Sivaram, a pro-LTTE journalist, and Joseph Pararajasingham, a pro-LTTE MP allegedly by the government of Sri Lanka.

At the beginning of 2008, the focus of the civil war turned to civilian targets, with commuter bus and train bombings carried out in most parts of the country, including a series of attacks against commuters in and around Colombo.

Talks and Further Violence

In light of this violence, the co-chairs of the Tokyo Donor conference called on both parties to return to the negotiating table. The co-chairs—the United States in particular—were heavily critical of the violence perpetrated by the LTTE. US State Department officials, as well as the US ambassador to Sri Lanka, gave warnings to the Tigers claiming a return to hostilities would mean that the Tigers would face a "more capable and more determined" Sri Lankan military. While the talks were going on there was violence targeted towards civilians such as massacre of 5 Tamil students on January 2, 2006, in Trincomalee when high school students playing by the beach were briefly detained and then shot dead.

In a last-minute effort to salvage an agreement between the parties, the Norwegian special envoy Erik Solheim and the LTTE theoretician Anton Balasingham arrived in the island. The parties severely disagreed on the location of the talks; however, continued efforts produced a breakthrough when both parties agreed on February 7, 2006, that new talks could be held in Geneva, Switzerland on February 22 and February 23. These talks were reported to have gone "above expectations", with both the government and the LTTE agreeing to curb the violence and to hold further talks on April 19–21.

During the weeks after the talks, there was a significant decrease in violence. However, the LTTE resumed attacks against the military in April beginning with a Claymore anti-personnel mine attack on military vehicles which killed 10 navy sailors on April 11. The following day, coordinated bombings by rebels and rioting in the north-eastern part of the country left 16 dead. First, a Claymore anti-personnel mine exploded in Trincomalee, killing two policemen in their vehicle. Another blast, set off in a crowded vegetable market, killed one soldier and some civilians. Ensuing rioting by civilians left more than a dozen dead. Responsibility for these attacks was claimed by an organisation called the Upsurging People's Force, which the military accused of being a front for the LTTE.

In light of this violence, the LTTE called for a postponement of the Geneva talks until April 24–25, and the government initially agreed to this. Following negotiations, both the government and the rebels agreed to have a civilian vessel transport the regional LTTE leaders with international truce monitors on April 16, which involved crossing government-controlled territory. However, the climate shifted drastically when the Tamil Tigers cancelled the meeting, claiming not to have agreed to a naval escort. According to the SLMM, the Tamil rebels had previously agreed to the escort. This led to Helen Olafsdottir, spokesperson for the SLMM saying "It was part of the agreement. The rebels should have read the clauses carefully. We are frustrated. "

On April 20, 2006, the LTTE officially pulled out of peace talks indefinitely. While they stated that transportation issues had prevented them from meeting their regional leaders, some analysts and the international community held a deep skepticism, seeing the transportation issue as a delaying tactic by the LTTE in order to avoid attending peace talks in Geneva.

Violence continued to spiral and on April 23, 2006, six Sinhalese rice farmers were massacred in their paddy fields by suspected LTTE cadres in the Trincomalee district. The following day, two suspected Tamil Tiger rebels were shot dead in Batticaloa when caught planting mines after rebels reportedly hacked a young mother to death and kidnapped her infant.

After LTTE launched a suicide assault on a naval convoy in which 18 sailors died, the Allaipiddy massacre of May 13, 2006, happened in which 13 minority Tamil civilians were killed in separate incidents in three villages in the islet of Kayts (Uruthota) in northern Sri Lanka. International condemnation against the LTTE skyrocketed following the attempted assassination of the commander of the Sri Lanka Army, Lieutenant General Sarath Fonseka by a pregnant LTTE Black Tiger suicide bomber Anoja Kugenthirasah, who blew herself up at the Sri Lankan Army headquarters in the capital, Colombo. Lt. Gen. Fonseka and twenty-seven others were injured, while ten people were killed in the attack. For the first time since the 2001 ceasefire, the Sri Lanka Air Force carried out aerial assaults on rebel positions in the north-eastern part of the island nation in retaliation for the attack.

This attack, along with the assassination of Lakshman Kadiragamar a year earlier and an unsuccessful attack against a naval vessel carrying 710 unarmed security force personnel on holiday, proved the catalysts as the European Union decided to proscribe the LTTE as a terrorist organisation on May 19, 2006. It resulted in the freezing of LTTE assets in the member nations of the EU, and put an end to its efforts to raise funds its terror campaign in Sri Lanka. In a statement, the European Parliament said that the LTTE did not represent all the Tamils and called on it to "allow for political pluralism and alternate democratic voices in the northern and eastern parts of Sri Lanka".

As the North and East of the country continued to be rocked by attacks, new talks were scheduled in Oslo, Norway, between June 8–9. Delegations from both sides arrived in Oslo, but the talks were cancelled when the LTTE refused to meet directly with the government delegation claiming its fighters were not been allowed safe passage to travel to the talks. Norwegian mediator Erik Solheim told journalists that the LTTE should take direct responsibility for the collapse of the talks.

Further violence followed, including the Vankalai massacre in which family of four minority Sri Lankan Tamils from the village of Vankalai in the district of Mannar in Sri Lanka on June 8, 2006, were tortured and killed. Both the mother and the nine year old daughter were allegedly raped before being killed. The gory images of the corpses published by the pro rebel Tamilnet news site created controversy in Sri Lanka and abroad. The Sri Lankan army and Tamil Tiger rebels have blamed each other for the killings. There was also the Kebithigollewa massacre on June 15, 2006, in which the LTTE attacked a bus killing at least 64 Sinhalese civilians and prompting more air strikes by the Air Force, and the assassination of Sri Lanka's third highest-ranking army officer and Deputy Chief of Staff General Parami Kulatunga on June 26 by an LTTE suicide bomber. These events led the SLMM to question whether a ceasefire could still be said to exist. However most analysts continued to believe that the return to full-scale war was unlikely and the "low-intensity conflict" would continue.

Mavil Oya (Mavil Aru) Water Dispute

A new crisis leading to the first large-scale fighting since signing of the ceasefire occurred when the LTTE closed the sluice gates of the Mavil Oya (Mavil Aru) reservoir on July 21 and cut the water supply to 15,000 villages in government controlled areas. After initial negotiations and efforts by the SLMM to open the gates failed, the Air Force attacked LTTE positions on July 26, and ground troops began an operation to open the gate.

The sluice gates were eventually reopened on August 8, with conflicting reports as to who actually opened them. Initially, the SLMM claimed that they managed to persuade the LTTE to lift the waterway blockade conditionally. However a government spokesman said that "utilities could not be used as bargaining tools" by the rebels and government forces launched fresh attacks on LTTE positions around the reservoir. These attacks prompted condemnation from SLMM Chief of Staff, who stated "(The government) have the information that the LTTE has made this offer. " "It is quite obvious they are not interested in water. They are interested in something else. " The LTTE then claimed they opened the sluice gates "on humanitarian grounds" although this was disputed by military correspondents, who stated the water began flowing immediately after the security forces carried out a precise bombing of the Mavil Aru anicut. Eventually, following heavy fighting with the rebels, government troops gained full control of the Mavil Aru reservoir on August 15.

LTTE Offensives in Muttur and Jaffna

As fierce fighting was ongoing in the vicinity of (Mavil Oya) Mavil Aru, the violence spread to Gokanna (Trincomalee), where the LTTE launched an attack on a crucial Sri Lanka Navy base, and to the strategic government controlled coastal town of Muttur in early August, resulting in the deaths of at least 30 civilians and displacing 25,000 residents of the area. The clashes erupted on August 2, 2006 when the LTTE launched a heavy artillery attack on Muttur and then moved in, gaining control of some parts of the town. The military retaliated, and re-established full control over the town by August 5, killing over 150 LTTE cadres in heavy fighting.

Soon afterwards, 17 persons working for the International French charity Action Against Hunger (ACF) in Mooduthara (Muthur), were found executed. They were found lying face down on the floor of their office, with bullet wounds, still wearing their clearly marked T-shirts indicating they were international humanitarian workers. The murders prompted widespread international condemnation. The SLMM claimed that the government was behind the attack, but the government denied the allegation calling it "pathetic and biased", and stated that the SLMM had "no right to make such a statement because they are not professionals in autopsy or post-mortem. " An official investigation launched by the government with the aid of international forensic experts is currently ongoing.

Meanwhile, in the north of the country, some of the bloodiest fighting since 2001 took place after the LTTE launched massive attacks on Sri Lanka Army defence lines in the Jaffna peninsula on August 11. The LTTE used a force of 400 to 500 fighters in the attacks which consisted of land and amphibious assaults, and also fired a barrage of artillery at government positions, including the key military airbase at Paluyaala (Palaly). Initially, the Tigers broke through army defense lines around Muhamalai (Mahakanda), and advanced further north, but they were halted after 10 hours of fierce fighting. Isolated battles continued over the next few days, but the LTTE was forced to give up its offensive due to heavy casualties. The LTTE is estimated to have lost over 250 cadres in the operation, while 90 Sri Lankan soldiers and sailors were also killed.

Chencholai air Strike

As ground battles were ongoing in the North and the East of Sri Lanka, Sri Lanka Air Force carried out an air strike against a facility in the rebel held Mullaitivu area, killing a number of Tamil girls.

Although the LTTE claimed 61 girls were killed, the SLMM stated they were able to count just 19 bodies. The government stated that it was an LTTE training facility and that the children were LTTE child soldiers, although the LTTE claimed the victims were schoolgirls attending a course on first aid at an orphanage.

Attack on the Pakistani High Commissioner

On the same day, a convoy carrying the Pakistani High Commissioner to Sri Lanka Bashir Wali Mohamed was attacked by a claymore antipersonnel mine concealed within an auto rickshaw. The High Commissioner escaped unhurt, but seven people were killed and a further seventeen injured in the blast. No group claimed responsibility for the attack, but the Sri Lankan government blamed the LTTE. The Pakistani High Commissioner, Bashir Wali Mohamed, claimed that India was strongly believed to have carried it out, in order to intimidate Pakistan, which is one of the main suppliers of military equipment to the Sri Lankan government. Pakistan had promised one shipload of the wherewithal every 10 days in coming months, it was Pakistan's assurance of solid support which prompted Sri Lankan Defence Secretary Gotabhaya Rajapakse to publicly state that Kilinochchi, the headquarters of the LTTE, would be liberated by the end of December.

Fall of Sampur

Since the resumption of violence, concerns were mounting among the military establishment that the strategically crucial Sri Lanka Navy base in Trincomalee was under grave threat from LTTE gun positions located in and around Sampur, which lies across the Koddiyar Bay from Trincomalee. Artillery fired from LTTE bases in the area could potentially cripple the naval base, bringing it to a complete standstill and therefore cutting the only military supply chain to Jaffna. All movements of naval vessels were also under the constant surveillance of the LTTE. These fears were backed up by a United States military advisory team which visited the island in 2005.

Following the clashes in Mavil Aru (Mavil Oya) and Muttur (Mooduthara), the LTTE had intensified attacks targeting the naval base in Trincomalee (Gokanna), and in a speech on August 21, Sri Lankan president Mahinda Rajapakse made clear the government intentions were to neutralize the LTTE threat from Sampur. On August 28, the Sri Lankan military launched an assault to retake the LTTE camps in Sampur and the adjoining Kaddaiparichchan (Gaeta-bara-hena) and Thoppur (Thupapura) areas. This led the LTTE to declare that if the offensive continued, the ceasefire would be officially over.

After steady progress, Sri Lankan security forces led by Brigade Commander Sarath Wijesinghe re-captured Sampur (Somapura) from the LTTE on September 4, and began to establish military bases there, as the LTTE admitted defeat and stated their cadres "withdrew" from the strategically important town. It marked the first significant territorial change of hands since the signing of the ceasefire agreement in 2002. The Sri Lankan Military estimated that 33 personnel were killed in the offensive, along with over 200 LTTE cadres.

LTTE Retaliation and Further Peace Talks

The LTTE struck back in October. First, they killed nearly 130 soldiers in a fierce battle at Muhamalai (Mahakanda), the crossing-point between government and LTTE controlled area in the north of the country. Just days later, a suspected LTTE suicide bomber struck a naval convoy in Habaraba, in the center of the country killing about 100 sailors who were returning home on leave. It was the deadliest suicide attack in the history of the conflict.

Two days later, LTTE Sea Tiger cadres launched an attack against the *Dakshina* naval base in the southern port city of Galle. It was the farthest south any major LTTE attack had taken place, and involved 15 LTTE cadres who arrived in five suicide boats. The attack was repulsed by the government, and the damage to the naval base was minimum. All 15 LTTE suicide cadres are believed to have died in the attack, along with one Sri Lanka Navy sailor.

Despite these incidents, both parties agreed to unconditionally attend peace talks in Geneva on October 28–29. However the peace talks broke down due to disagreements over the reopening of the key A9 highway, which is the link between Jaffna and government controlled areas in the south. While the LTTE wanted the highway, which was closed following fierce battles in August, to be reopened, the government refused, stating the LTTE would use it to collect tax from people passing through and would use it to launch further offensives against government troops.

Following the dawn of the new year, suspected LTTE cadres carried out two bus bombings in the south of the country, killing 21 civilians. News reports stated that the attacks bore all the hallmarks of an LTTE attack. The Sri Lankan government condemned the attacks and blamed the LTTE for carrying them out, although the LTTE denied any involvement. Iqbal Athas, an analyst for Jane's Defence Weekly commented that the LTTE's targeting of civilians was a cause for concern, and that further attacks against civilians couldn't be ruled out. Other analysts too expressed fears that LTTE attacks, which had largely been confined to military and political targets during the ceasefire period, may now increasingly target civilians as in earlier stages of a conflict.

Government Offensive in the East

In December 2006, the Commander of the Army and other senior government officials expressed their plans to initially drive the LTTE out of the Eastern Province of Sri Lanka, and then use the full strength of the military to defeat the LTTE in the North of the country. Among the reasons cited by the military for the offensives in the East were the need to "free the civilians in the area from the LTTE", who the military stated was firing artillery towards civilian settlements and were using 35,000 people as human shields. These claims were later backed by the civilians who told reporters that they were held by force by the Tamil Tigers. On November 7, 2006, in the midst of conflicting claims over 45 Tamil civilians were killed in what is known as the Vaharai bombing.

Subsequently, the Army began an offensive against the LTTE on December 8, 2006, in the Batticoloa district with the objective of taking Vakarai, the principle stronghold of the LTTE in the East, but temporarily aborted it after a week of fighting due to the large number of civilians in the area and the difficulty in conducting combat operations due to the ongoing Monsoon rain. Over the next few weeks, an estimated 20,000 civilians fled from Vaakare to Government controlled areas fearing the imminent assault. The Army launched a new offensive in mid January, and Vaakarr fell to the advancing troops on January 19, 2007. While the offensive in the East was ongoing, the LTTE and others accused the government of murdering 15 civilians in the Padahuthurai bombing on January 2, 2007, when the Sri Lanka Air Force bombed what they claimed to be rebel LTTE naval base in Illuppaikadavai in Northern Sri Lanka. The Army launched assaults from three different directions, and the LTTE and Defence Spokesman Keheliya Rambukwella announced that "The people of Vaakare have been liberated from the clutches of the terrorists". The loss of Vaakare (Vakarai) has been predicted to cut off supply routes of the northern Tigers to their cadres in the East, thus weakening the Tigers' already diminishing grip on the East.

As the military offensive was ongoing, the LTTE continued to carry out attacks against civilians in government held territory. On April 1, 2007, the Sri Lankan military accused the LTTE of killing six Sinhalese tsunami aid workers in the Eastern district of Batticaloa. The next day, suspected LTTE cadres set off a bomb aboard a civilian bus in Ampara which killed seventeen people, including three children.

Troops mostly operating in small groups of Special Forces and Commando units began a new operation in February to clear the last remaining LTTE cadres from the Eastern Province. As part of the operation, troops captured the key LTTE base in Gokatugolla (Kokkadicholai) on March 28, and the strategic A5 highway on April 12, bringing the entire highway under government control for the first time in 15 years. This meant the LTTE's presence in the East was reduced to a 140 square kilometre pocket of jungle land in the Thoppigala area north-west of Madakalapuva (Batticaloa). The offensive had left nine soldiers dead along with 184 Tiger cadres, with no civilian casualties, according to military estimates.

Government Offensive in the North. Targeting LTTE Leadership

Sporadic fighting in the North had been going on for months, but the intensity of the clashes increased after September 2007. During clashes in the Forward Defence Lines, separating their forces, both sides exchanged heavy artillery fire, after which military incursions followed. By December 22, 2007, the LTTE defences at Uyilankulama and Thampanai were lost to advancing troops of the Sri Lanka Army. On December 29, 2007, the Army overran the LTTE stronghold at Parappakandal, in Mannar District.

In an interview with the *Sunday Observer* the Army Commander Lt. Gen. Sarath Fonseka said that the Army had occupied the LTTE's Forward Defence Lines and surrounded the Wanni LTTE bases from all directions. He also said that there were around 3,000 Tigers remaining and that the military intended to annihilate them within the first six months of the next year. A day later there were less optimistic statements by Army, Air Force and Navy Commanders. The Army was to face an estimated 5,000 Tiger cadres in the Wanni. The Commander of the Army intended to shift the current battles in the Forward Defence Lines to a decisive phase in August 2008. In the Commanders' view, it was quite possible to defeat the LTTE in 2008.

The military of Sri Lanka claimed that the leader of the LTTE, Velupillai Prabhakaran, was seriously injured during air strikes carried out by the Sri Lanka Air Force on a bunker complex in Jayanthinagar on November 26, 2007. Earlier, on November 2, 2007, S. P. Thamilselvan, who was the head of the rebels' political wing, was killed during another government air raid. The Sri Lanka Air Force openly vowed to destroy the entire leadership of the LTTE. On January 5, 2008, Colonel Charles,

Head of LTTE Military Intelligence, was killed in a claymore mine ambush by a suspected Sri Lanka Army Deep Penetration Unit, according to a pro-LTTE website.

Timeline after Government's Withdrawal from Ceasefire

On January 2, 2008, the Sri Lankan government officially pulled out of the Ceasefire Agreement. This was amidst the demands of the defense secretary Gotabhaya Rajapaksa on December 29, 2007.

Donor countries such as the United States, Canada, and Norway have shown deep regrets on this decision by the Sri Lankan government. Neighboring India has also shown its dismay of Sri Lanka's abrogation of the ceasefire.

On January 10, 2008, the LTTE formally responded with the statement from the new head of the Tigers' political wing, B. Nadesan. According to him, the LTTE was shocked and disappointed, since the Government of Sri Lanka had unilaterally withdrawn from the ceasefire agreement without any justification.

It was further stated that even then the LTTE was ready to implement every clause of the CFA agreement and respect it one hundred percent. The LTTE claimed that, taking into account the acts of the government, the international community ought to immediately remove the bans it had placed on the LTTE.

It was reported that 185 Sri Lankan soldiers were killed on April 23, 2008, when troops manning the Muhamalai Forward Defence Line attempted to open a third front and advance towards Kilinochchi from Jaffna peninsula. The incident was a setback to military efforts to eliminate the rebels.

On May 9, 2008, the town of Adampan was captured by the Sri Lankan Army. On June 30, 2008, SLA troops linked up the Mannar battlefront with the Vavuniya battlefront, in the southwest of Periyamadhu.

On July 16, 2008, SLA troops captured Vidattaltivu, the biggest town situated on the North-Western coast of the island and the main Sea Tiger base. On July 20, 2008, the Sri Lankan Army captured the town of Iluppaikkadavai.

On July 21, 2008, the LTTE announced that it would be declaring a unilateral ceasefire from July 28 to August 4, to coincide with the 15th summit of the heads of state of SAARC to be held in Colombo. However, the government of Sri Lanka dismissed the LTTE's offer as needless and treacherous.

Significant Military Gains by the Government

On August 2, 2008, Vellankulam town, the last LTTE's stronghold in Mannar District, fell to the advancing SLA troops. This marked the liberation of the entire Mannar district by the Army, which took eight months. Two days earlier the Army crossed the Mannar-Kilinochchi boundary and entered the district of Kilinochchi. The defence secretary, Gotabhaya Rajapaksa, commented that the army was on target to take the rebel stronghold of Kilinochchi before the end of the year. Following weeks of heavy military confrontation, on September 2, 2008, the Army took complete control of the town of Mallavi.

On September 9, 2008, the LTTE launched a surprise attack on the Vavuniya air base. The Army claimed that the assault was repulsed, with heavy casualties on both sides, whereas the rebels claimed that the operation was a success.

On September 15, 2008, a fierce battle started in the area of Akkarayankulam, which is located in the proximity of Kilinochchi. On October 3, 2008, a UN aid convoy, consisting of 51 trucks carrying around 650 tonnes of food, managed to unload all its cargo in Kilinochchi District, whereas Kilinochchi town was reported to have been left by "nearly everyone".

On October 6, 2008, retired Major General Janaka Perera was killed in a suicide blast along with 26 other victims. The government blamed the attack on the LTTE. Meanwhile, the head of the army said his troops were within 2 km (1. 2 mi) of the Tigers' administrative headquarters in Kilinochchi.

On October 17, 2008, SLA troops cut off the Mannar-Poonaryn (A-32) road north of Nachchikuda, thus effectively encircling Nachchikuda, which was the main remaining Sea Tiger stronghold on the northwestern coast of the Island. At that point the situation of more than 200,000 civilians who had been displaced in the latest round of fighting was turning into a humanitarian disaster; however, due to a number of reasons including doubts regarding the sincerity of the LTTE's negotiations, neither Western governments nor India intervened to broker a new ceasefire.

On October 28, 2008, SLA troops on the Western Kilinochchi battlefront started a final assault on the LTTE's western coastal bastion at Nachchikuda, which fell the next day. After that the Army Task Force 1 continued their advance towards Pooneryn and captured Kiranchchi, Palavi, Veravil, Valaipadu and Devil's Point. On November 15, 2008, troops of the Army Task Force 1 entered the strategically important Tamil Tiger stronghold of Pooneryn.

On November 3, 2008, the newly created Army Task Force 3 was introduced into the area of Mankulam, with the objective of engaging the LTTE cadres in a new battlefront towards the east of the Jaffna–Kandy (A-9) road. SLA troops captured Mankulam and the surrounding area on November 17, 2008.

On the Eastern Mullaitivu battlefront SLA troops entered Alampil, 10 km (6. 2 mi) south of Mullaitivu, on December 4, 2008.

Fall of Kilinochchi and Subsequent Events

According to the Sri Lankan Army, an attack on Kilinochchi started on November 23, 2008. Troops were attacking rebels' defences from three directions. However, the LTTE offered a stiff resistance, and the prolonged attack resulted in heavy casualties on both sides. Not until January 1, 2009, were SLA troops able to capture Paranthan, which is located to the north of Kilinochchi along the A-9 route. According to unnamed defence observers, the fall of Paranthan isolated the southern periphery of the Elephant Pass LTTE foothold and also exposed the LTTE's main fortification at Kilinochchi.

On January 2, 2009, the President of Sri Lanka, Mahinda Rajapaksa, announced that troops had captured Kilinochchi, which the rebels had used for over a decade as their de facto administrative capital. It was noted that the loss of Killinochchi had caused a substantial dent in the LTTE's image as a capable, ruthless terrorist group. The fall of Kilinochchi to the Sri Lankan troops was celebrated by fire crackers throughout the island. It was also stated that after the fall of Kilinochchi the LTTE was likely to collapse before long under unbearable military pressure on multiple fronts.

As of January 8, 2009, the Tigers were abandoning their positions on the Jaffna peninsula to make a last stand in the jungles of Mullaitivu, their last main base. The entire Jaffna peninsula was captured by the Sri Lanka Army by January 14, 2009.

On January 25, 2009, SLA troops captured Mullaittivu town, the last Tamil Tiger rebel stronghold.

On February 3, 2009, the United States, the European Union, Japan and Norway issued a joint statement urging the Tamil Tigers to lay down their arms and end hostilities, as there was just a short time before the Tigers lost all the territory still under their control.

On February 5, 2009, the military captured the last Sea Tiger base in Chalai, reducing the territory under rebel control to less than some 200 km.

On February 20, 2009, two LTTE planes on a suicide mission attacked the Sri Lankan capital Colombo, killing 2 and wounding 45. Both planes were reportedly shot down by the Sri Lankan Air Force before they could damage the intended targets which were the Army Headquarters and the main Airforce base.

This stage of the war has been marked by increased brutality against civilians and rapidly mounting civilian casualties. On February 19, 2009, Human Rights Watch issued a report accusing the Sri Lankan army of "slaughtering" the civilians during indiscriminate artillery attacks (including repeated shelling of hospitals) and calling on the Sri Lankan government to end its policy of "detaining displaced persons" in military-controlled internment camps. Human Rights Watch also urged the Tamil Tigers to permit trapped civilians to leave the war zone and to "stop shooting at those who try to flee". The UN was also concerned over the condition of internally displaced persons and estimated that some 200,000 people were being squeezed into a narrow 14 square kilometre patch of land on the coast in Vanni, which the government had declared the 'no-fire zone'.

On March 26, 2009, the military claimed that there was only one square kilometre left in Tamil Tiger control outside the no-fire zone. Less than three years ago, the LTTE controlled 15,000 km^2. Political pressure was placed on Mahinda Rajapaksa to find a political solution to the conflict and he called for a meeting with parliamentarians allied with the Tigers, but they refused until the government resolved the humanitarian crisis faced by civilians trapped in the fighting.

Fighting in the 'No-fire Zone'

On April 5, 2009, SLA troops gained total control over the entire area of Puthukkudiyirippu, pushing the Tamil Tigers into the no-fire zone set up for civilians.

On April 20, 2009, the Sri Lankan army captured the 3-kilometre (2 mi) long earth bund, built by the LTTE on the western border of the no-fire zone, and rescued over 30,000 civilians who had been trapped at the Puthumathalan and Amplalavanpokkani areas. Meanwhile, the LTTE accused the Sri Lankan government of killing about 1,000 civilians and injuring many more during its latest military offensive. However, LTTE has attacked the fleeing civilians using a suicide bomber and killing around 15 civilians.

On April 21, 2009, Sri Lankan troops reportedly launched a 'final assault' against the LTTE, especially targeting its leader, Vellupillai Prabhakaran. At the same time, a mass Tamil exodus from the 'no-fire zone' was under way.

On April 22, 2009, two senior LTTE members (LTTE media co-ordinator Velayuthan Thayanithi, alias Daya Master, and a top interpreter Kumar Pancharathnam, alias George) surrendered to the advancing Sri Lankan army. This came as "a rude shock" and a major setback for the rebel leadership.

By April 25, 2009, the area under the LTTE was reduced to 10 km^2. While the Tamil exodus from the 'no-fire zone' continued, the UN estimated that around 6,500 civilians may have been killed and another 14,000 wounded between January 2009 and April 2009. According to the BBC, the land recaptured by the army from the rebels in the past few months is totally depopulated and the destruction is seen everywhere.

On April 29, 2009, the Ministry of Defence aired an interview called the Hidden Reality, with Daya Master and George, of their experience and time in the 'no-fire zone'. When asked why they had surrendered, both men stressed that rebels were shooting at the civilians and preventing them from escaping from the 'no-fire zone' to safety in government-controlled areas. They also confirmed that the LTTE were still abducting and conscripting children as young as 14 years old, and would fire at anyone who tried to resist. BBC news also reported that the Tigers are forcefully recruiting young children and are stopping people from moving out to government-controlled areas.

On May 8, 2009, a group of independent United Nations experts called on the Human Rights Council to urgently set up an international inquiry to address the "critical" situation in Sri Lanka amid fighting between the Army and Tamil rebels. According to the UN Office for the Coordination of Humanitarian Affairs (OCHA), over 196,000 people fled the conflict zone, a shrinking pocket of land on the north-east coastline, where clashes continued between government troops and the LTTE, while at least 50,000 people were still trapped there.

Dr. V. Shanmugarajah, working in the makeshift hospital in the conflict zone, made a claim (which the BBC stated was "impossible to verify") that at least 378 civilians were killed and another 1,122 hurt during heavy and lengthy shelling overnight on May 9, 2009. The official stated that the shelling came from territory controlled by the Sri Lankan army. The Ministry of Health denied that there was a government official with the BBC source's name in the conflict zone. The Sri Lankan military denied shelling the 'no-fire zone', and attributed it to the LTTE. Videos were released to the media of the site and the site was clearly affected by the war. A UN spokesman in Colombo, Gordon Weiss, said more than 100 children died during the "large-scale killing of civilians" and described the situation in northern

Sri Lanka as a "bloodbath". UN Secretary-General Ban Ki-moon said he was appalled at the killing of hundreds of Sri Lankan civilians caught in the middle of hostilities between the army and separatist Tamil rebels over the weekend. He voiced deep concern over the continued use of heavy weapons in the conflict zone, but also stressed that the "reckless disrespect shown by the LTTE for the safety of civilians has led to thousands of people remaining trapped in the area".

On May 13, 2009, Dr. T. Varatharajah in Mullivaikal told the BBC more than 50 people had died when two shells hit the zone's main hospital compound. Sources in the UN said they agreed with that figure and that 100 had been injured. A Sri Lankan technician working for the International Committee of the Red Cross was killed in shelling along with his mother, the organisation said. US President Barack Obama urged Sri Lanka to stop "indiscriminate shelling" of civilians and urged Tamil Tiger rebels to lay down their arms. In a unanimous statement the UN Security Council "strongly condemned the LTTE for its continued use of civilians as human shields and acknowledged the legitimate right of the Government of Sri Lanka to combat terrorism". At the same time, Council members "expressed deep concern at the reports of continued use of heavy calibre weapons in areas with high concentrations of civilians, and expected the Government of Sri Lanka to fulfil its commitment in this regard".

On May 16, 2009, Sri Lankan troops broke through LTTE defences and captured the last section of coastline held by Tamil Tiger rebels. The army reported it was set to "clear" remaining rebel-held land within days. Later the military claimed, allegedly citing intercepted LTTE communication, that rebels are preparing for a mass suicide after being effectively cut-off of escape routes. Some rebels have been reported to be blowing themselves up.

The Times newspaper has reported, claiming a source within U. N., that as many as 20,000 civilians were killed in the Safe Zone. The paper further asserted that some of the deaths were caused by the Tamil Tigers but most were as a result of shelling by the Sri Lankan military. The UN had estimated that 6,500 civilians had been killed in the three months to the middle of April, while it has no official figure after that date. *The Times* number asserts that the death toll soared to 1,000 each day in the final two weeks of the war. The UN says it has no confirmed estimates of civilian casualties and the Sri Lankan government has denied the *Times'* allegations. *The Guardian* newspaper, quoting another U. N. official, called the Times' figure as a "dangerous extrapolation".

The five doctors (Sivapalan, V. Shanmugaraja, Thurairaja Vartharaja, Sathyamoorthy and Ilancheliyan) who reported mass civilian casualties in the final phase of the civil war issued statements on 8 July 2009 recanting their reports. They said the causulaty figures they released were exaggerated and were handed to them by the Tamil Tigers. They stated that between January 2009 and the end of the war in May 2009 some 600-700 civilians were killed and twice that number had been injured. This contradicts the figures issued by international aid agencies: the United Nations says that 6,500 were killed between late January and early April; and the Red Cross evacuated 14,000 sick and injured people between mid-February and mid-May. The doctors also said that it was not true that a hospital had been shelled on 2 February. However, the UN and the Red Cross, both of whom had their own staff in the Safe Zone, say that the hospital was shelled and civilians were killed.

The United Nations says it stands by its casualty figures. Amnesty International says the doctors' recantation lacks credibility because they have been detained for two months by the Sri Lankan military without access to lawyers; they face torture, ill-treatment and other forms of duress; senior government officials have threatened the doctors with serious charges, including treason; and because the doctors' recent statements contradict independently verified facts.

Amnesty says it remains concerned about the doctors' safety. It has repeatedly called for an independent inquiry into alleged war crimes committed by both sides. It has also urged all independent organisations, including the UN and Red Cross, who collected information about the final phase of the war to release that information so that the world knows the truth about casualty figures and war crimes.

End of the War

May 16: President Declares Victory

Sri Lankan President Mahinda Rajapaksa declared victory on May 16, 2009. However, the war did not end until the following day. Sri Lankan troops raced to clear the last LTTE pockets of resistance. As the last LTTE strongpoints crumbled, Sri Lankan troops killed 70 rebels attempting to escape by boat. The whereabouts of LTTE leader Vellupillai Prabhakaran and other major rebel leaders were not certain, however, Sri Lanka's government announced that Prabhakaran was dead on May 17, 2009. Several other important LTTE commanders committed suicide. Upon hearing news of the Sri Lankan victory, people celebrated in Colombo.

May 17: Tigers Admit Defeat

The LTTE finally admitted defeat on May 17, 2009, with the rebels' chief of international relations, Selvarasa Pathmanathan, stating on the website that "This battle has reached its bitter end... We have decided to silence our guns. Our only regrets are for the lives lost and that we could not hold out for longer".

May 18: Prabhakaran is Dead

The Sri Lankan army forces confirmed that the leader of the LTTE, Velupillai Prabhakaran, was killed in the morning of May 18, 2009. The announcement on state television came shortly after the military said it had surrounded Prabhakaran in a tiny patch of jungle in the north-east. *The Daily Telegraph* wrote that, according to Sri Lankan TV, Prabhakaran was "... killed in a rocket-propelled grenade attack as he tried to escape the war zone in an Ambulance with his closest aides. Colonel Soosai, the leader of his "Sea Tigers" navy, and Pottu Amman, his intelligence chief were also killed in the attack. "

The head of the Sri Lankan army, General Sarath Fonseka, said the military had defeated the rebels and "liberated the entire country". Military spokesman Brigadier Udaya Nanayakkara confirmed Prabhakaran's death, saying 250 Tamil Tigers were also killed overnight. They were all hiding and fighting from within a zone that was designated for civilians only.

General Sarath Fonseka on May 19, 2009, confirmed that the body of Prabhakaran, had been found. The LTTE chief's body was found by the 53 Division troops led by Major General Kamal Goonarathne.

Reaction

Sri Lanka: United National Party leader Ranil Wickremasinghe, through a telephone call, congratulated President Mahinda Rajapaksa and the state's security forces on 18 May 2009 for their victory over the LTTE. In a press release issued that day, Roman Catholic Archbishop Oswald Gomis said:

I congratulate His Excellency President Mahinda Rajapakse, President of the Democratic Socialist Republic of Sri Lanka, for his very courageous leadership and thank the Chiefs of the Defense outfit who supported him with deep commitment and self-sacrifice. I also offer my deepest sympathies to those who laid down their lives in battle and those innocent civilians killed, trapped in war.

In a sense we could say that we have won the battle but the war is not ended. The war would end only on the day that we grow in nationhood realizing that we are all one people in one country with equal right. We have to realize the fact that we are a multi-ethnic, multi-religious and multi-cultural community. As such we are now left with the great task of nation-building forgetting our ethnic, political and religious differences.

It is imperative that there be a political formula that will inspire confidence and promote a sense of belonging among the minority groups in the country. We have to leave the sad and bitter memories of the past three decades and look positively and optimistically towards the future in hope. All of us have to share the blame for our division and forgive each other. We should have the humility and wisdom to learn from the sad experiences of that past. It is then, and only then, that we could build nationhood that will bring true peace and prosperity to our beloved country-Sri Lanka. Let us always remember that united we will flourish but divided we will perish.

With the announced end of the war, Sri Lanka's stock exchange registered its sixth highest percentage gain ever.

International Organisations

United Nations-At a press conference in Geneva on 19 May 2009, United Nations Secretary General Ban Ki-moon said, "I am relieved by the conclusion of the military operation, but I am deeply troubled by the loss of so many civilian lives. The task now facing the people of Sri Lanka is immense and requires all hands. It is most important that every effort be undertaken to begin a process of healing and national reconciliation. I listened very carefully to what President Rajapaksa said in his address to Parliament today. The legitimate concerns and aspirations of the Tamil people and other minorities must be fully addressed. " The Secretary General went on to announce his upcoming visit to the wartorn region.

European Union-The European Council met in Brussels on 18 May 2009, during which it adopted a statement calling on "the Government of Sri Lanka urgently to proceed towards a comprehensive political process" and "the President of Sri Lanka to outline a clear process leading to a fully inclusive political solution, based on consent, equality and the rule of law". The Council stated that such moves are the only way toward long-term security, post-conflict reconstruction and prosperity in Sri Lanka. The statement concluded: "The EU continues to call for appropriate action by the United Nations Human Rights Council. " The Times reported that EU member nations sold arms to the Sri Lankan government in spite of fears of human rights abuses.

Other Countries

Canada-Minister of Foreign Affairs Lawrence Cannon stated "Canadians are very concerned about the aftermath of the military action in Sri Lanka and the appalling effect it has had on civilians. This terrible, decades-long war has inflicted untold devastation and heartbreak on Sri Lankans. The Government of Canada wishes to express its concerns about civilian casualties, and to convey its condolences to the people of Sri Lanka and those around the world who have lost friends and family members in this horrific conflict.

Canada urges the Government of Sri Lanka to begin to find a long-term political solution that responds to the legitimate aspirations of all the people of Sri Lanka. Canada is prepared to assist Sri Lankan efforts to find political reconciliation and a lasting peace. "

India-The Indian Ministry of External Affairs issued a statement on 18 May 2009 saying "In a telephone conversation with External Affairs Minister Shri Pranab Mukherjee earlier today, the President of Sri Lanka confirmed that armed resistance by the LTTE has come to an end and that LTTE leader Velupillai Prabhakaran is dead.

India will work with the people and Government of Sri Lanka to provide relief to those affected by the tragic conflict, and to rapidly rehabilitate all those who have been displaced, bringing their lives to normalcy as soon as possible.

It is our view that as the conventional conflict in Sri Lanka comes to an end, this is the moment when the root causes of conflict in Sri Lanka can be addressed. This would include political steps towards the effective devolution of power within the Sri Lankan Constitution so that Sri Lankans of all communities, including the Tamils, can feel at home and lead lives of dignity of their own free will. "

Iran-Foreign Minister of the Islamic Republic of Iran Manouchehr Mottaki telephoned Foreign Minister Rohitha Bogollagama on 19 May 2009 to extend his warm congratulations to the President, the Foreign Minister and the Government of Sri Lanka on the success achieved by Sri Lanka in defeating LTTE terrorism.

"Iran has maintained close relations with Sri Lanka and has always condemned terrorism and, consistently upheld the sovereignty and the territorial integrity of Sri Lanka". He states that he was speaking on behalf of the President of Iran who has conveyed his best wishes to Sri Lanka and will personally speak to the President of Sri Lanka at a mutually convenient time. Iran has also offered assistance through the Red Crescent, for the emergency relief operations for the IDPs in the North.

Japan-The Japanese premier's office released a statement of Prime Minister Taro Aso's telephone conversation with Sri Lankan President Mahinda Rajapakse. The statement said the Prime Minister "welcomed the end of the civil war between the Sri Lankan government and the Liberation Tigers of Tamil Eelam", and that "it is now important to help internally displaced people and their resettlement as well as to start showing improvement in the political process towards peace-building". It also outlined that poverty was one of the factors that create fertile ground for terrorism and said he would like to see Sri Lanka build infrastructure, adding that Japan would support Sri Lanka's efforts as much as possible.

Maldives-Both President Mohamed Nasheed and Vice-president Mohammed Waheed Hassan congratulated the government and people of Sri Lanka for their tremendous success in effecting an end to the decades-old conflict in their country.

The President reaffirmed continued their support and solidarity with the government and people of Sri Lanka, as Sri Lanka celebrates as a unified nation. "I take this opportunity to express on behalf of the Government and the people of Maldives our sincerest best wishes to Your Excellency and the people of Sri Lanka".

"This momentous occasion in Sri Lanka's history will pave the way towards realising greater equality and justice for all Sri Lankans."

Norway-Norwegian Foreign Minister Jonas Gahr Store said in a statement "our thoughts go to all who have lost relatives and loved ones in the war. We must cooperate to aid the victims. People in the refugee camps must quickly be allowed to return home. "

Store also said the situation in the refugee camps for internally displaced people must be improved, in line with demands made by the United Nations.

Pakistan-Minister of State for Foreign Affairs of Pakistan, Nawabzada Malik Amad Khan telephoned Deputy Minister of Foreign Affairs of Sri Lanka, Hussein A. Bhaila congratulating Sri Lanka's "great victory over terrorism".

The Pakistani State Minister stated that Pakistan has always been a steadfast friend of Sri Lanka and strongly supported the country's unity, sovereignty and territorial integrity and re-affirmed his government's continued cooperation with Sri Lanka in countering terrorism. He requested the Deputy Minister to convey his good wishes and felicitations to President Mahinda Rajapaksa and Foreign Minister, Rohitha Bogollagama.

Philippines-A Department of Foreign Affairs press release, dated 22 May 2009: "welcomes the return of law and order in northern Sri Lanka and supports the Government of Sri Lanka's search for a comprehensive, fair, and lasting political solution to the problems faced by its Tamil minority.

The Philippines hopes that a lasting political solution will be crafted in order that the Tamil minority share in the fruits of peace in their country. "

Russia-"The government of Russia has extended warmest congratulations to the president and the government of Sri Lanka on the success achieved by the island nation in defeating LTTE (Liberation Tigers of Tamil Eelam) terrorism," the Department of Government Information said in a statement.

Officials from Sri Lanka's Foreign Ministry said Russia has said that it supports the fight of the Sri Lankan government against terrorism and separatism.

Russia hoped that the end to the bloody armed conflict that lasted in Sri Lanka for more than a quarter century will be a guarantee of the establishment of an enduring peace, security and stability in the country.

South Africa-The South African government released a statement through Deputy International Relations and Cooperation Minister Ehrahim Ebrahim regarding the end of military operations in Sri Lanka.

"The South African government expresses grave regret at the manner in which the military offensive was conducted and urges the United Nations Human Rights Commission to urgently investigate possible violations of international human rights law and contraventions of the Geneva Convention. ", said Ebrahim. The statement also called for immediate humanitarian aid and international media access to the areas affected. "The South African government has noted the conciliatory tone in the speech of President Mahinda Rajapaksa on 19 May 2009 and express our hope that the end of the military campaign will result in a peaceful dialogue with all minorities to address their long standing grievances. We will continue to support any efforts aimed at bringing about peace and reconciliation between the parties"

Singapore-The Singapore Ministry of Foreign Affairs issued the following statement 29 May 2009 after Minister for Foreign Affairs George Yeo met Sri Lankan Minister of Foreign Affairs Rohitha Bogollagama at the Shangri-La Dialogue held in Singapore: "Singapore is relieved to see an end to the long-standing conflict in Sri Lanka. The conflict had taken a great toll on the country. Not only have tens of thousands of lives been lost, but hundreds of thousands of Sri Lankan civilians have also been displaced from their homes. The final cessation of military operations by the Sri Lankan Government provides a short window of opportunity to close a sad chapter of history and quickly begin a process of genuine national healing and reconciliation. A long term agreement taking into full consideration the interests of all communities within Sri Lanka must be forged and implemented to ensure a lasting peace. "

George Yeo congratulated Rohitha Bogollagama and the Sri Lankan government over the victory of LTTE terrorists.

Switzerland-In a press statement dated 19 May 2009, Switzerland's government welcomed the end of the armed conflict. However, Switzerland "regrets that international humanitarian law has been violated and appeals to all parties to comply with and to ensure respect for international regulations and obligations in all circumstances."

"Switzerland calls on all parties to refrain from incitement to hatred and to work towards reconciliation by means of unilateral or jointly agreed measures. All parties and groupings as well as members of the diaspora should work openly and in conjunction with international institutions to initiate a reconciliation process and a sustainable solution in the framework of a political dialogue. "

Turkey-The president of Sri Lanka, Mahinda Rajapaksa, telephoned the president of Turkey, Abdullah Gul. The conversation was released by the Turkish Presidency Media Center. The Sri Lankan president informed Gul about the recent achievements in the fight against terrorism. President Gul stated that he was pleased with the recent events and Turkey was ready to provide humanitarian aid, in the telephone conversation.

United Kingdom-Foreign Secretary David Miliband made the following written statement to the House of Commons on 19 May 2009: "On 19 May, the Sri Lankan President formally announced that on 18 May military forces had retaken all the territory once held by the LTTE and that they had captured or killed the senior leadership of that organisation. Many Sri Lankans of all communities, Sinhalese, Tamil and Muslim, will be relieved that the long and brutal conflict may at last be over. Sri Lanka has before it an historic opportunity to resolve the underlying causes of the conflict and ensure a lasting peace. We must continue to work with Sri Lanka's Government and all its communities to ensure that this opportunity is taken and that it leads to a sustainable end to the conflict. The continuing focus of this Government's activity over the coming days and weeks, will be to work with international partners in encouraging the Sri Lankan Government to devote as much energy to winning the peace as it did to winning the war. "

Lord Malloch Brown, the Minister of State Foreign and Commonwealth office in UK said, in the House of Lords, "Indeed, in our initial contacts with the president, we congratulated him on finishing of a brutal 26-year war, which was instigated by the Tamil Tigers-a terrorist group" replying to Lord Naseby who raised the matter in the House of Lords. Agreeing with Lord Naseby, Lord Malloch Brown said, "the political solution to this must come from inside Sri Lanka from a process set up and led by President Rajapaksa. "

Speaking further, Brown said, "But we also made it extremely clear to him that, whether or not that victory would be seen as the opening of a new and happier chapter in Sri Lanka depended on whether he could now go that next step and show the statesmanship to find a political as well as humanitarian solution to this community's issues. "

He further said, "On 17 May, the Prime Minister announced an additional œ5 million in humanitarian aid for Sri Lanka, taking the total to œ12. 5 million since September 2008. " Speaking before him Lord Naseby said, "My Lords, have Her Majesty's Government congratulated the Sri Lankan Government on defeating the Tamil Tigers and bringing peace to the country? On the international front, is it Her Majesty's Government's policy primarily to tackle the resettlement of the 250,000 Tamils and the 100,000 Muslims who were ethnically cleansed from Jaffna, or is it to continue to lecture that there should be a constitutional settlement, which really rests with the Parliament of Sri Lanka?"

United States-At a press conference in Washington, D. C., on 18 May 2009, United States Department of State spokesman Ian Kelly said: "The Department of State welcomes the fact that the fighting has ended, and we are relieved that the immense loss of life and killing of innocent civilians appears to be over. This is an opportunity for Sri Lanka to turn the page on its past and build a Sri Lanka rooted in democracy, tolerance, and respect for human rights. Now is the time for the government to engage the Tamils, Sinhalese, and other Sri Lankans to create a political arrangement that promotes and protects the rights of all Sri Lankans. It is also vital for the government to provide for the needs of the 280,000 civilians now living in relief camps. Providing food, water, shelter, basic health care, and sanitation, as well as expediting their return to their homes should be a top priority for the government. "

The Sri Lankan defence ministry said that US Ambassador to Sri Lanka Robert O. Blake, Jr., called Foreign Minister Rohitha Bogollagama on 18 May 2009 regarding humanitarian aid to displaced persons and reconciliation with the Tamil people.

Vietnam-On 21 May 2009, Foreign Affairs Ministry spokesman Le Dung released the following statement in response to journalist inquiries:

> *"Vietnam welcomes the recent victory of the Government and people of Sri Lanka. This victory will create favourable conditions for Sri Lanka to concentrate on the cause of national construction and development, contributing to peace, stability and development in the region. "*

Combat after May 18th 2009

19 May 2009: 3 LTTE cadre killed by the Sri Lankan Army at Kachikudichchiaru, Ampara.

20 May 2009: 5 LTTE cadre killed by the Sri Lankan Army near Periyapillumalai area.

21 May 2009: 10 LTTE cadre killed by the Sri Lankan Army in the Kadawana jungle area.

27 May 2009: 11 LTTE cadre killed by the Sri Lankan Army at Kalavanchchikudi in the Batticaloa area

Five T-56 assault rifles, twenty claymore mines (15 kg each), two hand grenades, three anti-personnel mines and medical items were reported recovered by military sources.

5 June 2009: Special Task Force (STF) personnel while conducting search and clear operation in Darampalawa area in Ampara confronted with a group of LTTE cadre and recovered two bodies along with numerous military items.

5 August 2009: Selvarasa Pathmanathan, the new leader of LTTE had been arrested in a South East Asian country and brought to Colombo.

Aftermath

President Rajapaksa also said the government was committed to a political solution, and for this purpose action would be taken based on the 13th Amendment to the Constitution.

The continuous defeats of the LTTE had made its cadres abandon the outfit in large numbers. Over 7,237 militants who had surrendered to the military are now being rehabilitated at various rehabilitation centres. Among them are 1,601 females.

The officials said over 2,379 LTTE militants were identified by the military. Among them are militants who volunteered to surrender to the troops.

Over 2,065 males said that they surrendered to the troops at the Forward Defence Lines as they were disillusioned with their leadership. They had sensed the historic defeat and realised that the LTTE cadres had to fight to protect the life of the LTTE leader and not for Eelam. The militants who surrendered are provided with facilities and most of them have been sent to rehabilitation camps.

Since the end of the war, more than 5,000 Tamil youths have been gathering at selected police stations in Eastern Province for interviews to join the police force as the government has called for interviews. The Sri Lankan government is planning to recruit 2,000 new police officers to the department, especially for the services in the northern region of the country.

Humanitarian Impact

Towards the end of the war, as Sri Lankan government forces advanced deeper into Tamil Tiger controlled areas, international concern grew for the fate of the 350,000 civilians trapped. On 21 January 2009, the Sri Lankan military declared a 32 square kilometers (12. 4 mi) *Safe Zone* located northwest of Puthukkudiviruppu, between the A35 Highway and the Chalai Lagoon. Sri Lankan Air Force aircraft dropped leaflets urging civilians to relocate to the safe zone and wait until the army could move them into safer locations. The Sri Lankan military promised not to fire into the area. However, only small numbers of civilians actually crossed into the Safe Zone, and the Sri Lankan government, the United Nations, and Human rights organizations accused the LTTE of preventing civilians from leaving. The fighting eventually caused civilians to flee the safe zone to a narrow strip of land between Nanthi Kadal and the Indian Ocean. The Sri Lankan military declared a new 10 square kilometre (3. 9 square mile) Safe Zone northwest of Mullativuon February 12. Over the next three months, the Sri Lankan military repeatedly attacked the Safe Zone with aircraft and artillery to destroy the last remnants of the Tamil Tigers trapped there. The Sri Lankan government claimed that it was trying to hit Tamil Tiger positions, and claimed that these raids started on February 15 and ended on April 19, the day before the Army breached Tamil Tiger defenses, and civilians started to pour out. However, these attacks caused heavy damage.. Thousands of civilians were killed or injured, and the Tamil Tigers reportedly held many as human shields. The final stages of the war created 300,000 internally displaced persons

(IDPs) who were transferred to camps in Vavuniya District and detained there against their will. This, together with the conditions inside the camps, has attracted much criticism from inside and outside Sri Lanka. The camps were surrounded by barbed wire, and internees attempting to escape were shot. The government has stated that it will release civilians from the camps once it completes a screening process to weed out Tamil Tigers hiding among civilians in the camps, as well as demining operations. On 7 May 2009 the Sri Lankan government announced plans to resettle 80% of the IDPs by the end of 2009. After the end of the civil war President Rajapaksa gave assurances to foreign diplomats that the bulk of the IDPs would be resettled in accordance with the 180 day plan. By 9 October 2009, after more than 150 days, less than 10% (27,000) of the IDPs had been released or returned to their places of origin, with 250,000 still being held in the camps.

Protests

Tamil diaspora communities around the world have recently been actively protesting the Tamil civilian casualties in Northern Province, Sri Lanka and the war in general. Active protests have occurred in the major and capital cities of India, Britain, Canada, Australia, Norway, Switzerland, Denmark, Germany and the United States. The collective objective of the protests was to persuade world national leaders to stop the civil war and bring a permanent ceasefire with an internationally coordinated diplomatic strategy.

Casualties

The Sri Lankan civil war was very costly, killing an estimated 80,000-100,000 people. The deaths include 27,639 Tamil fighters, more than 23,327 Sri Lankan soldiers and policemen, 1,155 Indian soldiers, and tens of thousands of civilians.

Secretary of Defence Ministry Gotabhaya Rajapaksa said on an interview with state television that 23,790 Sri Lankan military personnel were killed since 1981.

From the August 2006 recapture of the Mavil Aru reservoir until the formal declaration of the cessation of hostilities (on May 18), 6261 Sri Lankan soldiers were killed and 29,551 were wounded.

The Sri Lankan military estimates that up to 22,000 LTTE militants were killed in the last three years of the conflict.

The final five months of the civil war saw the heaviest civilian casualties. The UN, based on credible witness evidence from aid agencies as well civilians evacuated from the *Safe Zone* by sea, estimated that 6,500 civilians were killed and another 14,000 injured between mid-January 2009, when the *Safe Zone* was first declared, and mid-April 2009. There are no official casualty figures after this period but estimates of the death toll for the final four months of the civil war (mid-January to mid-May) range from 15,000 to 20,000. A US State Department report has suggested that the actual casualty figures were probably much higher than the UN's estimates and that significant numbers of casualties weren't recorded. A former UN official has claimed that up to 40,000 civilians may have been killed in the final stages of the civil war.

The Nagaland Problem

Various groups are involved in the Insurgency in Northeast India, India's north east states, which are connected to the rest of India by a narrow strip of land known as the Siliguri Corridor or Chicken's Neck. Much of the region is notably ethnically and linguistically different from the rest of India. In the region several armed factions operate. Some groups call for a separate state, others for autonomy while some extreme groups demand nothing but complete independence.

Northeastern India consists 7 states (also known as *the seven sisters*): Assam, Meghalaya, Tripura, Arunachal Pradesh, Mizoram, Manipur, and Nagaland. Tensions exist between these states and the central government as well as amongst the tribal people, who are natives of these states, and migrant peoples from other parts of India.

The states have accused New Delhi of ignoring the issues concerning them. It is this feeling which has led the natives of these states to seek greater participation in self-governance. There are existing territorial disputes between Manipur and Nagaland.

There is a rise of insurgent activities and regional movements in the northeast, especially in the states of Assam, Nagaland, Mizoram and Tripura. Most of these organizations demand independent state status or increased regional autonomy and sovereignty. Regional tensions have eased off as of late, with Indian and state governments' concerted effort to raise the living standards of the people in these regions. However, militancy still exists within the region. Among the rebellions in the area are Tripura Rebellion and Assam Conflict.

Kashmir Conflict

The Kashmir conflict refers to the territorial dispute over Kashmir, the north western most region of South Asia. The parties to the dispute are India, Pakistan, China, and the people of Kashmir.

India claims the entire former princely state of Jammu and Kashmir and presently administers approximately 43% of the region including most of Jammu, Kashmir Valley, Ladakh and the Siachen Glacier. India's claim is contested by Pakistan which controls approximately 37% of Kashmir, mainly Azad Kashmir and the northern areas of Gilgit and Baltistan. In addition, China controls 20% of Kashmir including Aksai Chin which it occupied following the brief Sino-Indian War of 1962 and the Trans-Karakoram Tract, also known as the Shaksam Valley, that was ceded to it by Pakistan in 1963.

India's official position is that Kashmir is an integral part of India. Pakistan's official position is that Kashmir is a disputed territory whose final status must be determined by the people of Kashmir. China's official position is that Aksai Chin is a part of Tibet, which is a part of China. Certain Kashmiri independence groups believe that Kashmir should be independent of both India and Pakistan.

India and Pakistan have fought three wars over Kashmir: in 1947, 1965, and 1999. India and China have clashed once, in 1962 over Aksai Chin as well as the northeastern Indian state of Arunachal Pradesh. India and Pakistan have also been involved in several skirmishes over the Siachen Glacier.

Since after 1987, disputed State elections resulted in some of the states legislative assembly to form militant wings later on after the election, creating the catalyst for the insurgency,; the Indian-administered Jammu and Kashmir has been the site of conflict between the Indian Armed Forces, militants and separatists. Furthermore, India alleges these militants are supported by Pakistan. This turmoil in Jammu and Kashmir has resulted in thousands of deaths., however the insurgency over the past two decades has died down according to the Indian government. On the other hand, there have been protest movements in Indian Administered Kashmir since 1989. The movements were created to voice Kashmir's disputes and grievances with the Indian government, specifically the Indian Military. Elections held in 2008 were generally regarded as fair by the United Nations High Commissioner for Refugees, had a high voter turnout in spite of calls by militants for a boycott, and led to pro India Jammu & Kashmir National Conference forming the government in the state. According to Voice of America, many analysts have interpreted the high voter turnout in this election as a sign that the people of Kashmir have endorsed Indian rule in the state.

A 2001 report 'Pakistan's Role in the Kashmir Insurgency' of US Think tank RAND corporation noted that 'More intrinsically, the nature of the Kashmir conflict has been transformed from what was originally a secular, locally-based struggle (conducted via the Jammu Kashmir Liberation Front-JKLF) to one that is now largely carried out by foreign militants and rationalized in pan-Islamic religious terms. ' Most of the militant organizations are composed of foreign mercenaries mostly from the Pakistani Punjab In 2010 with the support of its intelligence agencies Pakistan has been once again 'boosting' Kashmir militants and recruitment of 'martyrs' in Pakistani state of Punjab has increased.

Timeline

Partition and Dispute

From 1820 until the Partition of India in 1947, Kashmir was governed by the Maharaja of Kashmir who were Hindu, although the majority of the population were Muslim, except in the Jammu region. On partition Pakistan expected Kashmir to be annexed to it.

In 1947, British rule in India ended with the creation of two new nations: the Union of India and the Dominion of Pakistan while British suzerainty over the 562 Indian princely states ended. According to the Indian Independence Act 1947, "the suzerainty of His Majesty over the Indian States lapses, and with it, all treaties and agreements in force at the date of the passing of this Act between His Majesty and the rulers of Indian States", so the states were left to choose whether to join India or Pakistan or to remain independent. Jammu and Kashmir, the largest of the princely states, had a predominantly Muslim population while having a hindu ruler (Maharaja Hari Singh.)

In October 1947, Muslim revolutionaries in western Kashmir and Pakistani tribals from Dir entered Kashmir intending to liberate it from Dogra rule. Unable to withstand the invasion, the Maharaja signed the Instrument of Accession that was accepted by the government of India on 27 October 1947.

Indo-Pakistani War of 1947

After rumours that the Maharaja was for the union with India, Muslim revolutionaries from western Kashmir and Pakistani tribals made rapid advances into the Baramulla sector. Maharaja Hari Singh of Kashmir asked the government of India to intervene, However, India and Pakistan had signed an agreement of non-intervention (maintenance of the status quo in Jammu and Kashmir.) Although tribal fighters from Pakistan had entered Jammu and Kashmir, there was no iron-clad legal evidence to unequivocally prove that Pakistan was officially involved. It would have been illegal for India to unilaterally intervene in an open, official capacity unless Jammu and Kashmir officially joined the Union of India, at which point it would be possible to send in its forces and occupy the remaining parts.

The Maharaja desperately needed military assistance when the Pathan tribals reached the outskirts of Srinagar. Before their arrival into Srinagar, India argues that Maharaja Hari Singh completed negotiations for acceding Jammu and Kashmir to India in exchange for receiving military aid. The agreement which ceded Jammu and Kashmir to India was signed by the Maharaja and Lord Mountbatten of Burma.

The resulting war over Kashmir, the First Kashmir War, lasted until 1948, when India moved the issue to the UN Security Council. The UN previously had passed resolutions setting up for the monitoring of the conflict in Kashmir. Following the set up of the UNCIP the UN Security Council passed Resolution 47 on 21 April 1948. The resolution imposed an immediate cease-fire and called on Pakistan to withdraw all military presence. In addition, the resolution also stated that Pakistan would have no say in Jammu and Kashmir politics. India would retain a minimum military presence and "the final disposition of the State of Jammu and Kashmir will be made in accordance with the will of the people expressed through the democratic method of a free and impartial plebiscite conducted under the auspices of the United Nations. " The cease fire was enacted on 31 December 1948.

At that time, the Indian and Pakistani governments agreed to hold the plebiscite but Pakistan did not withdraw its troops from Kashmir thus violating the condition for holding the plebiscite. Over the next several years, the UN Security Council passed four new resolutions, revising the terms of Resolution 47 to include a synchronous withdrawal of both Indian and Pakistani troops from the region, per the recommendations of General Andrew McNaughton. To this end, UN arbitrators put forward 11 different proposals for the demilitarization of the region-every one of which was accepted by Pakistan, but rejected by the Indian government. The resolutions were passed by United Nations Security Council under Chapter VI of the United Nations Charter. Resolutions passed under Chapter VI of UN charter are considered non binding and have no mandatory enforceability as opposed to the resolutions passed under Chapter VII.

Sino-Indian War

In 1962, troops from the People's Republic of China and India clashed in territory claimed by both. China won a swift victory in the war, resulting in the Chinese administration of the region called Aksai Chin, which continues to date. In addition to these lands, another smaller area, the Trans-Karakoram, was demarcated as the Line of Control (LOC) between China and Pakistan, although parts on the Chinese side are claimed by India to be parts of Kashmir. The line that separates India from China in this region is known as the *Line of Actual Control.*

1965 and 1971 Wars

In 1965 and 1971, heavy fighting again broke out between India and Pakistan. The Indo-Pakistani War of 1971 resulted in the defeat of Pakistan and Pakistan Military's surrender in East Pakistan (Bangladesh). The Simla Agreement was signed in 1972 between India and Pakistan. By this treaty, both countries agreed to settle all issues by peaceful means and mutual discussions in the framework of the UN Charter.

Militancy

In 1989, a widespread armed insurgency started in Kashmir, Since after the 1987 State election disputes resulted in some of the states legislative assembly forming militant wings later on after the election creating the catalyst for the Mujahadeen insurgency, which continues to this day. Timeline of the conflict. India contends that this was largely started by the large number of Afghan mujahadeen who entered the Kashmir valley following the end of the Soviet-Afghan War, though Pakistan and Kashmiri nationalists argue that Afghan mujahideen did not leave Afghanistan in large numbers until 1992, three years after the insurgency began. Yasin Malik, a leader of one faction of the Jammu Kashmir Liberation Front, along with Ashfaq Majid Wani and Farooq Ahmad Dar alias Bitta Karatay, was one of the Kashmiris to organize militancy in Kashmir. However since 1995, Malik has renounced the use of violence and calls for strictly peaceful methods to resolve the dispute. He developed differences with one of the senior leader, Farooq Papa, for shunning the demand for independent Kashmir and trying to cut a deal with the Indian Prime Minister resulting in spilt in which Bitta Karatay, Salim Nanhaji and other senior comrades joined Farooq Papa. Pakistan claims these insurgents are Jammu and Kashmir citizens, and are rising up against the Indian army in an independence movement. Pakistan also accuses the Indian army of committing serious human rights violations in Kashmir. Pakistan denies that it has or currently is supplying weapons and ammunition to the insurgents.

India claims these insurgents are Islamic terrorist groups from Pakistan-administered Kashmir and Afghanistan, fighting to make Jammu and Kashmir part of Pakistan. They claim Pakistan is supplying munitions to the terrorists, and training them in Pakistan. India also states that the terrorists have been killing many citizens in Kashmir, and committing human rights violations, while denying that its own armed forces are responsible for the human rights abuses. On a visit to Pakistan in 2006 current Chief Minister of Kashmir Omar Abdullah remarked that foreign militants, who had nothing to do with Kashmir, were engaged in reckless killings and mayhem in the name of religion. Indian government has said militancy is now on the decline.

The Pakistani government calls these insurgents, "Kashmiri freedom fighters", and claims that it gives only moral and diplomatic support to these insurgents, though India believes they are Pakistan-supported terrorists from Pakistan Administered Kashmir. In October 2008 President Asif Ali Zardari of Pakistan called the Kashmir separatists Terrorists in an interview with The Wall Street Journal, these comments by Zardari sparked outrage amongs many Kashmiris, some of whom defied a curfew by the Indian army to burn his effigy.

The peaceful protest movement has been a 'purely indigenous, purely Kashmiri (Quoted by Washington post from Mirwaiz Farooq a Kashmiri party leader) 'Gandhi style' (stated by Wall Street Journal) peaceful protest movement in Indian Administered Kashmir since 1989. The movement was created for the same reason as the insurgency ;the disputed rigged elections in 1987,Kashmir dispute and grievances with the Indian government specifically the Indian Military that has committed human rights violations. This reinforced by the United Nations that has said India has committed Human rights violations.

Al-Qaeda Involvement

In a 'Letter to American People' written by Osama bin Laden in 2002 he stated that one of the reasons he was fighting America is because of her support to India on the Kashmir issue. In 2002 U. S. Secretary of Defense Donald Rumsfeld on a trip to Delhi suggested that Al-Qaeda was active in Kashmir though he did not have any hard evidence. An investigation in 2002 unearthed evidence that Al-Qaeda and its affiliates were prospering in Pakistan-administered Kashmir with tacit approval of Pakistan's National Intelligence agency Inter-Services Intelligence In 2002 a special team of Special Air Service and Delta Force was sent into Indian Administered Kashmir to hunt for Osama bin Laden after reports that he was being sheltered by Kashmiri militant group Harkatul Mujahideen. U. S. officials believe that Al-Qaeda was helping organize a campaign of terror in Kashmir in order to provoke conflict between India and Pakistan. Their strategy was to force Pakistan to move its troops to the border with India thereby relieving pressure on Al-Qaeda elements hiding in northwestern Pakistan. U. S. Intelligence analysts say al-Qaeda and Taliban operatives in Pakistan-administered Kashmir are helping terrorists they had trained in Afghanistan to infiltrate Indian administered Kashmir. The leader of the Harkat-ul-Mujahideen a major Kashmiri militant group, Fazlur Rehman Khalil, signed al-Qaeda's 1998 declaration of holy war, which called on Muslims to attack all Americans and their allies. In 2006 Al-Qaeda claim they have established wing in Kashmir this has worried the Indian government.

However the Indian Army Lt. Gen. H. S. Panag, GOC-in-C Northern Command said to reporters that the army has ruled out the presence of Al Qaeda in Indian administered Jammu and Kashmir furthermore he said that there is nothing that can verify reports from the media of Al Qaeda presence in the state. He however stated that Alqaeda had strong ties with Kashmir militant groups Lashkar-e-Taiba and Jaish-e-Mohammed in Pakistan. In January 2010 U. S. Defense secretary Robert Gates while on a visit to Pakistan stated that Al-qaeda was seeking to destabilize the region and planning to provoke a nuclear war between India and Pakistan.

On September 2009 U. S. Drone strike reportedly killed Ilyas Kashmiri who was the chief of Harkat-ul-Jihad al-Islami a Kashmiri militant group associated with Al Qaeda. Kashmiri was described by Bruce Riedel as a 'prominent' Al-qaeda member while others have described him as head of military operations for Al-Qaeda. It was noted that Waziristan had now become the new battlefield for Kashmiri militants who were now fighting NATO in support of Al-Qaeda. Kashmiri was also charged by U. S. in a plot against Jyllands-Posten the Danish newspaper which was at the center of Jyllands-Posten Muhammad cartoons controversy.

Conflict in Kargil

In mid-1999 insurgents and Pakistani soldiers from Pakistani Kashmir infiltrated into Jammu and Kashmir. During the winter season, Indian forces regularly move down to lower altitudes as severe climatic conditions makes it almost impossible for them to guard the high peaks near the Line of Control. The insurgents took advantage of this and occupied vacant mountain peaks of the Kargil range overlooking the highway in Indian Kashmir, connecting Srinagar and Leh. By blocking the highway, they wanted to cut off the only link between the Kashmir Valley and Ladakh. This resulted in a high-scale conflict between the Indian Army and the Pakistan Army.

At the same time, fears of the Kargil War turning into a nuclear war provoked the then-US President Bill Clinton to pressure Pakistan to retreat. Faced with mounting losses of personnel and posts, Pakistan Army withdrew the remaining troops from the area ending the conflict. India reclaimed control of the peaks which they now patrol and monitor all year long.

Reasons Behind the Dispute

The Kashmir Conflict arises from the Partition of India in 1947 into modern India and Pakistan. Both the countries have made claims to Kashmir, based on historical developments and religious affiliations of the Kashmiri people. The state of Jammu and Kashmir, which lies strategically in the Northwest of the subcontinent, bordering China and the former Soviet Union, was a princely state ruled by Maharaja Hari Singh, under the paramountcy of British India. In geographical and legal terms, the Maharaja could have joined either of the two new Dominions. Although urged by the Viceroy, Lord Mountbatten of Burma, to determine the future of his state before the transfer of power took place, Hari Singh demurred. In October 1947, incursions and counter-incursions by Pakistan and India have taken place leading to a war, as a result of which the state of Jammu and Kashmir remains divided between the two countries.

Two-thirds of the former princely state (known as the Indian state of Jammu and Kashmir), comprising Jammu, the Kashmir Valley, and the sparsely populated Buddhist area of Ladakh are controlled by India; one-third is administered by Pakistan. The latter includes a narrow strip of land called Azad Kashmir and the Northern Areas compromising the Gilgit Agency, Baltistan and the former kingdoms of Hunza and Nagar. Attempts to resolve the dispute through political discussions were unsuccessful. In September 1965, war broke out again between Pakistan and India. The United Nations called for a yet another cease-fire, and peace was restored once again following the Tashkent Declaration in 1966, by which both nations returned to their original positions along the demarcated line. After the 1971 war and the creation of independent Bangladesh, under the terms of the 1972 Simla Agreement between Prime Minister Indira Gandhi of India and Zulfiqar Ali Bhutto of Pakistan, it was agreed that neither country would seek to alter the cease-fire line in Kashmir, which was renamed as the Line of Control, "unilaterally, irrespective of mutual differences and legal interpretations".

Numerous violations of the Line of Control have occurred, including the infamous incursions by insurgents and Pakistani armed forces at

Kargil leading to the Kargil war. There are also sporadic clashes on the Siachen Glacier, where the Line of Control is not demarcated and both countries maintain forces at altitudes rising to 20,000 ft (6,100 m).

Indian View

Indian viewpoint is succinctly summarized by Ministry of External affairs, Government of India —

- India holds that the Instrument of Accession of the State of Jammu and Kashmir to India, signed by the Maharaja Hari Singh (erstwhile ruler of the State) on 26 October 1947, was completely valid in terms of the Government of India Act (1935), Indian Independence Act (1947) and international law and was total and irrevocable.
- The Constituent assembly of Jammu and Kashmir had unanimously ratified the Maharaja's Instrument of Accession to India and had adopted a constitution for the state that called for a perpetual merger of the state with the Union of India. India claims that this body was a representative one, and that its views were those of the Kashmiri people at the time.
- United Nations Security Council Resolution 1172 tacitly accepts India's stand regarding all outstanding issues between India and Pakistan and urges the need to resolve the dispute through mutual dialogue and does not call for a plebiscite.
- United Nations Security Council Resolution 47 cannot be implemented since Pakistan failed to withdraw its forces from Kashmir which was the first step in implementing the resolution. Now the resolution is obsolete since the geography and demographics of the region have been permanently altered. The resolution was passed by United Nations Security Council under Chapter VI of the United Nations Charter. Resolutions passed under Chapter VI of UN charter are considered non binding and have no mandatory enforceability as opposed to the resolutions passed under Chapter VII.
- India does not accept the two-nation theory that forms the basis of Pakistan and argues that Kashmir, despite being a Muslim-majority state, is in many ways an "integral part" of secular India.
- The state of Jammu and Kashmir was provided significant autonomy in the Article 370 of the Constitution of India.
- All differences between India and Pakistan including Kashmir need to be settled through bilateral negotiations as agreed to

by the two countries when they signed the Simla Agreement on 2 July 1972.

Additional Indian viewpoint regarding the broader debate over the Kashmir conflict include:

- India believes that the insurgency and terrorism in Kashmir is deliberately being fuelled by Pakistan to create instability in the region. The Government of India has repeatedly asked the international community to declare Pakistan as a sponsor of terrorism.
- Pakistan is trying to raise anti-India sentiment among the people of Kashmir by spreading false propaganda against India. According to the state government of Jammu and Kashmir, Pakistani radio and television channels deliberately spread "hate and venom" against India to alter Kashmiri opinion.
- In a diverse country like India, disaffection and discontent are not uncommon. Indian democracy has the necessary resilience to accommodate genuine grievances within the framework of our sovereignty, unity and integrity. Government of India has expressed its willingness to accommodate the legitimate political demands of the people of the state of Jammu and Kashmir.
- India has asked United Nations that it should not be leave unchallenged or unaddressed claims of moral, political and diplomatic support for terrorism, which were clearly in contravention of United Nations Security Council Resolution 1373 which is a Chapter VII resolution that makes it mandatory for member states to not provide active or passive support to terrorist organizations. Specifically, it has pointed out that the Pakistani government continues to support various terrorist organizations, such as Jaish-e-Mohammad and Lashkar-e-Taiba, in direct violation of this resolution.
- India points out at various reports by human rights organizations condemning Pakistan for the lack civic liberties in Pakistan-administered Kashmir. According to India, most regions of Pakistani Kashmir, especially Northern Areas, continue to suffer from lack of political recognition, economic development and basic fundamental rights.

Pakistani View

Pakistan's claims to the disputed region are based on the rejection of Indian claims to Kashmir, namely the Instrument of Accession. Pakistan insists that the Maharaja was not a popular leader, and was regarded as a tyrant by most Kashmiris, Pakistan also maintains that the Maharaja used brute force to suppress the population. Pakistan also accuses India of hypocrisy, as it refused to recognize the accession of Junagadh to Pakistan and Hyderabad's independence, on the grounds that those two states had Hindu majorities (in fact, India occupied and forcibly integrated those two territories). Furthermore, as he had fled Kashmir due to Pakistani invasion, Pakistan asserts that the Maharaja held no authority in determining Kashmir's future. Additionally, Pakistan argues that even if the Maharaja had any authority in determining the plight of Kashmir, he signed the Instrument of Accession under duress, thus invalidating the legitimacy of his actions.

Pakistan also claims that Indian forces were in Kashmir before the Instrument of Accession was signed with India, and that therefore Indian troops were in Kashmir in violation of the Standstill Agreement, which was designed to maintain the status quo in Kashmir (although India was not signatory to the Agreement, signed between Pakistan and the Hindu ruler of Jammu and Kashmir).

From 1990 to 1999 some organizations report that Indian Armed Forces, its paramilitary groups, and counter-insurgent militias have been responsible for the deaths 4,501 of Kashmiri civilians. Also from 1990 to 1999, there are records of 4,242 women between the ages of 7-70 that have been raped. Similar allegations were also made by some human rights organizations.

In short, Pakistan holds that:

- The popular Kashmiri insurgency demonstrates that the Kashmiri people no longer wish to remain within India. Pakistan suggests that this means that either Kashmir wants to be with Pakistan or independent.
- According to the two-nation theory which is one of the theories that is cited for the partition that created India and Pakistan, Kashmir should have been with Pakistan, because it has a Muslim majority.
- India has shown disregard to the resolutions of the UN Security Council, and the United Nations Commission in India and Pakistan by failing to hold a plebiscite to determine the future allegiance of the entire state.
- The Kashmiri people have now been forced by the circumstances to rise against the alleged repression of the Indian army and uphold their right of self-determination through militancy. Pakistan claims to give the Kashmiri insurgents moral, ethical and military support.

- Recent protests in Indian administered Kashmir show a large number of people showing increased anger over Indian rule with massive rallies taking place to oppose Indian control of the state.
- Pakistan also points to the violence that accompanies elections in Indian Kashmir and the anti Indian sentiments expressed by some people in the state.
- Pakistan has noted the wide spread use of extrajudicial killings in Indian-administered Kashmir carried out by Indian security forces while claiming they were caught up in encounters with militants. These fake encounters are common place in Indian-administered Kashmir and the perpetrators are spared criminal prosecution. These fake encounters go largely uninvestigated by the authorities.
- Pakistan points towards reports from the United Nations which condemn India for its human rights violations against Kashmiri people.
- Pakistan President Asif Ali Zardari stated in October 2008 that Kashmiri 'freedom fighters' were terrorists. However his remarks met with widespread condemnation across Pakistan and Kashmir, including prominent politicians.
- The Chenab formula-This was proposed in 1960's, in which Kashmir valley and other Muslim dominated areas north of Chenab river will go to Pakistan, and Jammu and other Hindu dominated region will go to India.
- Pakistan points towards the numerous Human rights violations which occur within Indian-administered Kashmir and the many reports by Human rights organization strongly condemning Indian troops for widespread rape and murder of innocent civilians accusing these civilians of being killed in encounters

Chinese View

- Chinese never accepted the British negotiated boundary in the north east area of the princely state of Kashmir (Aksai Chin).
- The Chinese settled their border disputes in the region with Pakistan in 1963.

Cross-border Troubles

The border and the Line of Control separating Indian and Pakistani Kashmir passes through some exceptionally difficult terrain. The world's highest battleground, the Siachen Glacier is a part of this difficult-to-man boundary. Even with 200,000 military personnel, India maintains that it is infeasible to place enough men to guard all sections of the border throughout the various seasons of the year. Pakistan has indirectly acquiesced its role in failing to prevent "cross border terrorism" when it agreed to curb such activities after intense pressure from the Bush administration in mid 2002.

The Government of Pakistan has repeatedly claimed that by constructing a fence along the line of control, India is violating the Shimla Accord. However, India claims the construction of the fence has helped decrease armed infiltration into Indian-administered Kashmir.

In 2002 Pakistani President and Army Chief General Pervez Musharraf promised to check infiltration into Jammu and Kashmir.

Water Dispute

Another reason behind the dispute over Kashmir is water. Kashmir is the origin point for many rivers and tributaries of the Indus River basin. They include Jhelum and Chenab which primarily flow into Pakistan while other branches-the Ravi, Beas and the Sutlej irrigate northern India. Pakistan has been apprehensive that in a dire need, India (under whose portion of Kashmir lies the origins and passage of the said rivers) would use its strategic advantage and withhold the flow and thus choke the agrarian economy of Pakistan. The Boundary Award of 1947 meant that the headwaters of Pakistani irrigation systems were in Indian Territory. The Indus Waters Treaty signed in 1960 resolved most of these disputes over the sharing of water, calling for mutual cooperation in this regard. But this treaty faced issues raised by Pakistan over the construction of dams on the Indian side which limit water to the Pakistani side.

Human Rights Abuse

Indian Administered Kashmir

In Jammu and Kashmir, India, the violent Islamic insurgency has specifically targeted the Hindu Kashmiri Pandit minority, violated their human rights and 400,000 have either been murdered or displaced. US Congressman Frank Pallone stated "The conflict in Kashmir cannot be separated from the global war against terrorism, over the past fifteen years militant forces, including elements of Al-Qaeda and the Taliban, have used violence against the Kashmiri Pandits in an effort to institute Islamic rule in this region".

However the Indian Army Lt. Gen. H. S. Panag, GOC-in-C Northern Command said to reporters that the army has ruled out the presence of Al Qaeda in Jammu and Kashmir furthermore he said that there is nothing that can verify reports from the media of Al Qaeda presence in the state. This violence been condemned and labelled as ethnic cleansing in a 2006 resolution passed by the United States Congress. Also in 2009 Oregon Legislative Assembly passed a resolution to recognize 14 September 2007, as Martyrs Day to acknowledge ethnic cleansing and campaigns of terror inflicted on non-Muslim minorities of Jammu and Kashmir by militants seeking to establish an Islamic state. On the other hand the CIA reported about 300,000 Hindus in Indian Administered Kashmir valley are internally displaced due to militancy and about 100,000-50,000 Kashmiri Muslim refugees because of Islamic militants atrocities and Islamic militants fighting with Indian army. UNHCR reported that there are roughly 1. 5 million Refugees from Indian-administered Kashmir in Pakistan administered Kashmir and Pakistan.

Claims of human rights abuses have been made against the Indian Armed Forces and the armed militants operating in Jammu and Kashmir. A 2005 study conducted by Medecins Sans Frontieres found that Kashmiri women are among the worst sufferers of sexual violence in the world, with 11. 6% of respondents reporting that they had been victims of sexual abuse. Some surveys have found that in the Kashmir region itself (where the bulk of separatist and Indian military activity is concentrated), popular perception holds that the Indian Armed Forces are more to blame for human rights violations than the separatist groups. According to the MORI survey of 2002, in Kashmir only 2% of respondents believed that the militant groups were guilty of widespread human rights abuses, while 64% believed that Indian troops were guilty of the same. This trend was reversed however in other parts of the state. Off late Amnesty International has called on India to “unequivocally condemn enforced disappearances” and to ensure that impartial investigation is conducted on reality of mass graves in its controlled Kashmir region. As the Indian state police confirms as many as 331 deaths while in custody and 111 enforced disappearances since 1989... Amnesty again criticise Indian Military in an incident on 22 April, several armed forces personnel forcibly entered the house of a 32-year-old woman in the village of Wawoosa in the Rangreth district of Jammu and Kashmir.

They reportedly molested her 12-year-old daughter and raped her other three daughters, aged 14, 16 and 18. When another woman attempted to prevent soldiers from attacking her two daughters, she was beaten. Soldiers reportedly told her 17-year-old daughter to remove her clothes so that they could check whether she was hiding a gun. They molested her before leaving the house.

Several international agencies and the UN have reported human rights violations in Indian-administered Kashmir. In a recent press release the OHCHR spokesmen stated "The Office of the High Commissioner for Human Rights is concerned about the recent violent protests in Indian-administered Kashmir that have reportedly led to civilian casualties as well as restrictions to the right to freedom of assembly and expression. ". A 1996 Human Rights Watch report accuses the Indian military and Indian-government backed paramilitaries of "committ[ing] serious and widespread human rights violations in Kashmir.

" One such alleged massacre occurred on 6 January 1993 in the town of Sopore. *TIME Magazine* described the incident as such: "In retaliation for the killing of one soldier, paramilitary forces rampaged through Sopore's market setting buildings ablaze and shooting bystanders. The Indian government pronounced the event 'unfortunate' and claimed that an ammunition dump had been hit by gunfire, setting off fires that killed most of the victims. " In addition to this, there have been claims of disappearances by the police or the army in Kashmir by several human rights organizations. Jammu and Kashmir Public Safety Act, 1978: Human rights organizations have also asked Indian government to repeal the Public Safety Act, since "a detainee may be held in administrative detention for a maximum of two years without a court order. ".

Many human rights organizations such as Amnesty International and the Human Rights Watch (HRW) have condemned human rights abuses in Kashmir by Indians such as "extra-judicial executions", "disappearances", and torture; the "Armed Forces Special Powers Act", which "provides impunity for human rights abuses and fuels cycles of violence. The Armed Forces Special Powers Act (AFSPA) grants the military wide powers of arrest, the right to shoot to kill, and to occupy or destroy property in counterinsurgency operations. Indian officials claim that troops need such powers because the army is only deployed when national security is at serious risk from armed combatants. Such circumstances, they say, call for extraordinary measures. " Human rights organizations have also asked Indian government to repeal the Public Safety Act, since "a detainee may be held in administrative detention for a maximum of two years without a court order. ". A 2008 report by the United Nations High Commissioner for Refugees determined that Indian Administered Kashmir, was only 'partly Free'.

Pakistan Administered Kashmir

Pakistan as an Islamic Republic imposes multiple restrictions on peoples religious freedom in areas of Kashmir under its control. Shias and Ismailis are subject to discrimination and have been targets of sectarian violence. The majority of population of Northern Areas is Shia unlike rest of Pakistan which is majority Sunni. The constitution of Azad Kashmir specifically prohibits activities that may be prejudicial to the states accession to Pakistan and as such regularly suppresses demonstrations against the government. A number of Islamist militant groups operate in this area including Al-Qaeda with tacit permission from Pakistan's intelligence. As in Indian administered Kashmir there have been allegations of Human rights abuse in Pakistan administered Kashmir. The Balawaristan National Front has stated its goal of seeking independence from Pakistan. Abdul Hamid Khan Chairman of Balawaristan National Front states that 'The Pakistani administration has also been involved in efforts to alter the demographic profile of Pakistan-occupied Gilgit Baltistan, reducing the indigenous people to a minority. ' The Gilgit-Baltistan area is administered directly by Islamabad. The population here, primarily Shia Muslims, was brought under one federally administered territory administered by Pakistan on 16 November 1947, in the name of Islam. ' Other groups like Gilgit-Baltistan United Movement are demanding full autonomy for the areas of Gilgit and Baltistan. On 8 January 2005 11 people were killed following an armed attack on a Shia leader. A 2-day conference on Gilgit Baltistan was held on April 8–9, 2008 at the European Parliament in Brussels under the auspices of International Kashmir Alliance. Here several members of the European Parliament (MEPs) expressed concern over the human rights violation in Gilgit Baltistan and urged the government of Pakistan to establish democratic institutions and rule of law in this area of northern Kashmir. Abdul Hamid Khan, Chairman Balawaristan National Front speaking at the same conference said "no democratically elected representative (from Gilgit Baltistan) was included when Karachi Agreement was signed between Pakistan and Muslim Conference leaders in 1949. " According to Shaukat Ali chairman of International Kashmir alliance "On one hand Pakistan claims to be the champion of the right of self-determination of the Kashmiri people, but she has denied the same rights under its controlled parts of Kashmir and Gilgit Baltistan" Gilgit Baltistan region has been described a 'simmering cauldron of discontent. Continued deprivation of human rights has driven people to desperation. Many people think that they are under the colonial rule of Pakistani government and have boycotted independence day celebrations. Since independence Pakistan government has made no attempts to provide basic human rights in this region and no democratic setup exists.

A report 'Kashmir: Present Situation and Future Prospects' which was submitted to European Parliament by Emma Nicholson, Baroness Nicholson of Winterbourne was severely critical of lack of human rights, justice, democracy & Kashmiri representation in Pakistan National Assembly in Pakistan administered Kashmir. International Crisis Group has stated "Almost six decades after Pakistan's independence, the constitutional status of the Federally Administered Northern Areas (Gilgit and Baltistan), once part of the former princely state of Jammu and Kashmir and now under Pakistani control, remains undetermined, with political autonomy a distant dream. The region's inhabitants are embittered by Islamabad's unwillingness to devolve powers in real terms to its elected representatives, and a nationalist movement, which seeks independence, is gaining ground. The rise of sectarian extremism is an alarming consequence of this denial of basic political rights". However in 2009 Pakistan government implemented autonomy package for the people from Gilgit-Baltistan. This package was rejected as an eyewash by Balawaristan National Front whose spokesperson stated "It's meant to detract the international community from the violation of human rights in this region. " Manzoor Hussain Parwana, chairman Gilgit-Baltistan United Movement stated "The so-called provincial setup aims at concealing the human rights violations and continue the colonial control over the region. " The 'first step' being an election to elect their own assembly and there have been criticism about this move by Pakistan. Furthermore according to Reuters U. S many of the people from the region would rather join Pakistan as a province than integrated into Kashmir. Many people protested the elections and some people carried banners saying "Pakistan's expansionist designs in Gilgit-Baltistan are unacceptable" In December 2009 activists of nationalist Kashmiri groups staged a protest in Muzaffarabad to condemn rigging of elections and 'State Terrorism" by Pakistani forces.

According to Human Rights Commission of Pakistan Pakistan's Inter Services Intelligence operates in Pakistan administered Kashmir and is involved in extensive surveillance, arbitrary arrests, torture and murder. Generally this is done with impunity and perpetrators go unpunished. A 2008 report by United Nations High Commissioner for Refugees determined that Pakistan administered Kashmir was 'Not Free'.

Map Issues

As with other disputed territories, each government issues maps depicting their claims in Kashmir as part of their territory, regardless of actual control. It is illegal in India to exclude all or part of Kashmir in a map. It is also illegal in Pakistan not to include the state of Jammu and Kashmir as disputed territory, as permitted by the United Nations. Non-participants often use the Line of Control and the Line of Actual Control as the depicted boundaries, as is done in the CIA World Factbook, and the region is often marked out in hashmarks, although the Indian government strictly opposes such practices. When Microsoft released a map in Windows 95 and Map Point 2002, a controversy was raised because it did not show all of Kashmir as part of India as per Indian claim. However, all the neutral and Pakistani companies claim to follow UN's map and over 90% of all maps containing the territory of Kashmir show it as disputed territory.

Sources from: UN: The boundaries and names shown and the designations used on the map do not imply official endorsement or acceptance by the United Nations. Dotted line represents approximately the Line of Control of Jammu and Kashmir agreed upon by the Republic of India and the Government of Pakistan since 1972. Both the parties have not yet agreed upon the final status of the region and nothing significant has been implemented since the peace process began in 2004.

Islamabad: The Government of Pakistan maintains un-provisionally and unconditionally stating that the informal "Accession of Jammu and Kashmir" to Pakistan or even to the Republic of India remains to be decided by UN plebiscite. It accepts UN's map of the territory. Also the designations and the presentation of the Kashmir's regional map based on United Nations Organization practice, do not imply the expression of any opinion whatsoever on the part of the Commonwealth Secretariat or the publishers concerning the legal status of any country, territory or area, or of its authorities, or concerning the delimitation of its frontiers or boundaries. There is no intention to define the status Jammu and/or Kashmir, which has not yet been agreed upon by the parties.

New Delhi: The Government of India states that "the external artificial boundaries of India, especially concerning the Kashmir region under its jurisdiction created by a foreign body are neither correct nor authenticated".

Recent Developments

India continues to assert their sovereignty or rights over the entire region of Kashmir, while Pakistan maintains that it is a disputed territory. Pakistan argues that the status quo cannot be considered as a solution. Pakistan insists on a UN sponsored plebiscite. Unofficially, the Pakistani leadership has indicated that they would be willing to accept alternatives such as a demilitarized Kashmir, if sovereignty of

Azad Kashmir was to be extended over the Kashmir valley, or the 'Chenab' formula, by which India would retain parts of Kashmir on its side of the Chenab river, and Pakistan the other side-effectively re-partioning Kashmir on communal lines. The problem however is that the Population of Pakistan Administered portion of Kashmir is both ethnically and linguistically and culturally different from that in Kashmir Valley India. The Azad Kashmir population being on the most part ethnic Punjabis. T

herefore a Partition on the Chenab formula is opposed by most Kashmiri politicians from all spectrums, though some, such as Sajjad Lone, have in recent months suggested that non-Muslim part of Jammu and Kashmir be separated from Kashmir and handed to India. Some political analysts say that the Pakistan terrorist state policy shift and mellowing down of its aggressive stance may have to do with its total failure in the Kargil War and the subsequent 9/11 attacks that put pressure on Pakistan to alter its terrorist position. Further many neutral parties to the dispute have noted that UN resolution on Kashmir is no longer relevant. Even the European Union has viewed that the plebiscite is not in Kashmiris' interest. The report also notes, that the UN-laid down conditions for such a plebiscite have not been, and can no longer be, met by Pakistan. Even the Hurriyat Conference observed in 2003, that "Plebiscite no longer an option" Besides the popular factions that support either parties, there is a third faction which supports independence and withdrawal of both India and Pakistan. These have been the respective stands of the parties for long, and there have been no significant change over the years. As a result, all efforts to solve the conflict have been futile so far.

The Freedom in the World 2006 report categorized the Indian-administered Kashmir as "partly free", and Pakistan-administered Kashmir as well as the country of Pakistan "not free". India claims that contrary to popular belief, a large proportion of the Jammu and Kashmir populace wish to remain with India. A MORI survey found that within the Kashmir Valley, 9% of respondents said they felt they would be better off as Indian citizens, with 78% saying that they did not know, and the remaining 13% favouring Pakistani citizenship. According to a 2007 poll conducted by the Centre for the Study of Developing Societies in New Delhi, 87% of respondents in the Kashmir Valley prefer independence over union with India or Pakistan.

The 2005 Kashmir earthquake, which killed over 80,000 people, led to India and Pakistan finalizing negotiations for the opening of a road for disaster relief through Kashmir.

Efforts to end the Crisis

The 9/11 attacks on the US resulted in the US government wanting to restrain militancy in the world, including Pakistan. US urged Islamabad to cease infiltrations, which continue to this day, by Islamist militants into Indian-administered Kashmir. In December 2001, a terrorist attack on the Indian Parliament linked to Pakistan resulted in war threats, massive deployment and international fears of nuclear war in the subcontinent.

After intensive diplomatic efforts by other countries, India and Pakistan began to withdraw troops from the international border 10 June 2002, and negotiations began again. Effective 26 November 2003, India and Pakistan have agreed to maintain a ceasefire along the undisputed International Border, the disputed Line of Control, and the Siachen glacier. This is the first such "total ceasefire" declared by both nuclear powers in nearly 15 years. In February 2004, Pakistan further increased pressure on Pakistanis fighting in Indian-administered Kashmir to adhere to the ceasefire. The nuclear-armed neighbours also launched several other mutual confidence building measures. Restarting the bus service between the Indian-and Pakistani-administered Kashmir has helped defuse the tensions between the countries. Both India and Pakistan have also decided to cooperate on economic fronts.

On Dec. 5, 2006, Pakistani President Pervez Musharraf told an Indian TV channel that Pakistan would give up its claim on Kashmir if India accepted some of his peace proposals, including a phased withdrawal of troops, self-governance for locals, no changes in the borders of Kashmir, and a joint supervision mechanism involving India, Pakistan and Kashmir, the BBC reported. Musharraf also stated that he was ready to give up the United Nations' resolutions regarding Kashmir.

2008 Militant Attacks

In the week of 10 March 2008, 17 people were wounded when a blast hit the region's only highway overpass located near the Civil Secretariat– Indian-controlled Kashmir's seat of government– and the region's high court. A gun battle between security forces and militants fighting against Indian rule left five people dead and two others injured 23 March 2008. The battle began when security forces raided a house on the outskirts of the capital city of Srinagar. The Indian Army has been carrying out cordon-and-search operations against militants in Indian-administered Kashmir since the current armed violence broke out here in 1989. While the authorities here say 43,000 persons have been killed in the violence, various rights groups and non-governmental organizations have put the figure at twice that number.

According to Govt. of India Home Ministry, 2008 marks the lowest civilian casualties in 20 years with 89 deaths, compared to highest of 1,413 in 1996. 85 security personnel died in 2008 compared to 613 in 2001, while 102 militants killed. Human right situation improved with only 1 custodial death and no custodial disappearance.

2008 Kashmir Protests

Massive demonstrations occurred after plans by the Indian-administered Jammu and Kashmir state government to transfer 100 acres (0. 40 km) of land to a trust which runs the Hindu Amarnath shrine in the Muslim-majority Kashmir valley. This land was to be used to build a shelter to house Hindu pilgrims temporarily during their annual pilgrimage to the Amarnath temple.

Indian security forces and the Indian army responded quickly to keep order. More than 40 unarmed protesters were killed and at least 300 were detained. The largest protests saw more than a half million people waving Pakistani flags and crying for freedom at a single rally according to Time magazine. Pro-Independent Kashmir Leader Mirwaiz Umar Farooq warned that the peaceful uprising could lead to violent upsurge if India's heavy-handed crackdown on protests were not restrained. The United Nations expressed concern on India's response to peaceful protests and urged to investigate and bring to justice Indian security personnel who had taken part in the crackdown.

Separatists and workers of a political party were believed to be behind stone pelting incidents which led to retaliatory fire by the police. Autorickshaw laden with stones meant for distribution was seized by the police in March 2009. Furthermore, following the unrest in 2008, which included more than 500,000 protesters at a rally on 18 August, secessionist movements gained a boost.

2008 Kashmir Elections

State Elections were held in Indian held Kashmir in seven phases starting 17 November and finishing on 24 December 2008. In spite of calls by separatists for a boycott an unusually high turnout of almost 50% was recorded. The National Conference party which was founded by Sheikh Abdullah and regarded as pro India emerged with maximum seats and will form government in coalition with Indian National Congress.

2008 marks the lowest civilian casualties in 20 years with 89 deaths, compared to highest of 1,413 in 1996. 85 security personnel died in 2008 compared to 613 in 2001, while 102 militants were killed. Many analysts say Pakistan's preoccupation with jihadis within its own borders explains the relative calm.

2008 marked the greatest number of anti India protests since 1980 due to the Amarnath land transfer controversy with several hundred thousand protesters spilling out onto the streets of Indian-administered Kashmir demanding freedom from India the protests were suppressed by the Indian army with attacks on protesters leading to the deaths of 40 unarmed civilians. However the elections which were held subsequently led to almost half of the Kashmiris ignoring the boycott call by separatists and voting Pro India party National Conference into power. Separatists insist that this was so because people were looking towards their well being and voting for whatever could get them 'bread and clothing' and the turnout did not necessarily reflect the feelings of the Kashmiris towards India. On 30 December Congress and the National Conference agreed to form a coalition government, with Omar Abdullah as Chief Minister. On 5 January 2009 Omar Abdullah was sworn in as 11th Chief Minister of Jammu and Kashmir. In March 2009 Omar Abdullah stated that only 800 militants were active in the state and out of these only 30% were Kashmiris.

Obama on Kashmir Conflict

In an interview with Joe Klien of Time magazine in October 2008 Barack Obama expressed his intention to try to work with India and Pakistan to resolve this crisis in a serious way. He said he had talked to Bill Clinton about it (being a mediator). In an editorial in The Washington Times, Selig S Harrison, director of Asia Programme at the Center for International Policy and a senior scholar of the Woodrow Wilson International called it Obama's first foreign policy mistake. The Australian in an editorial called Obama's Idea to appoint a presidential negotiator "a very stupid and dangerous move indeed" In an editorial in Forbes, Reihan Salam associate editor for The Atlantic noted "The smartest thing President Obama could do on Kashmir is probably nothing. We have to hope that India and Pakistan can work out their differences on Kashmir on their own". The Boston Globe in an editorial called the idea of appointing Bill Clinton as an envoy to Kashmir "a mistake". India has long regarded Kashmir as an Integral part of India and resisted outside intervention considering Kashmir to be an integral part of India and the conflict a bilateral matter between India and Pakistan. President Obama appointed Richard Holbrooke as special envoy to Pakistan and Afghanistan. President Asif Ali Zardari had hoped that Holbrooke would help mediate to resolve Kashmir issue.

Subsequently Kashmir was removed from the mandate of Richard Holbrooke. "Eliminating ... Kashmir from his job description ... is seen as a significant diplomatic concession to India that reflects increasingly warm ties between the country and the United States," The Washington Post noted in a report. Brajesh Mishra, India's former national security adviser, was quoted in the same report as saying in reference to the territory's Indian-administered sector "No matter what government is in place, India is not going to relinquish control of Jammu and Kashmir," "That is written in stone and cannot be changed. " According to The Financial Times India has warned US President Barack Obama that he risks "barking up the wrong tree" if he seeks to broker a settlement between Pakistan and India over the disputed territory of Kashmir.

In July 2009 US Assistant Secretary of State Robert O. Blake, Jr. stated categorically that United States had no plans of appointing any special envoy to settle the long standing dispute of Kashmir between India and Pakistan calling it an issue which needs to be sorted out bilaterally by the two neighboring states. According to Dawn in Pakistan this will be interpreted as an endorsement of India's position on Kashmir that no outside power has any role in this dispute.

Chapter 2

Trade Route of the World

A trade route is a logistical network identified as a series of pathways and stoppages used for the commercial transport of cargo. Allowing goods to reach distant markets, a single trade route contains long distance arteries which may further be connected to several smaller networks of commercial and non commercial transportation. Historically, the period from 1532 BCE–1 CE saw the Western Asian, Mediterranean, Chinese and Indian societies develop major transportation networks for trade. Europe's early trading routes included the Amber Road, which served as a dependable network for long distance trade. Maritime trade along the Spice route became prominent during the Middle Ages; nations resorted to military means for control of this influential route. During the Middle Ages organizations such as the Hanseatic League, aimed at protecting interests of the merchants and trade, also became increasingly prominent. With the advent of modern times, commercial activity shifted from the major trade routes of the Old World to newer routes between modern nation states. This activity was sometimes carried out without traditional protection of trade and under international free trade agreements, which allowed commercial goods to cross borders with relaxed restrictions. Innovative transportation of the modern times includes pipeline transport, and the relatively well known trade using rail routes, automobiles and cargo airlines.

Development of Early Routes

Early Development

The period extending from the middle of the 2nd millennium BCE to the beginning of the Common Era saw the Western Asian, Mediterranean, Chinese and Indian societies develop major transportation networks for trade. One of the vital instruments which facilitated long distance trade was portage and the domestication of beasts of burden. Organized caravans, visible by the 2nd millennium BCE, could carry goods across a large distance as fodder was mostly available along the way. The domestication of camels allowed Arabian nomads to control the long distance trade in spices and silk from the Far East to the Arabian Peninsula. However, caravans were useful in long-distance trade largely for carrying luxury goods, the transportation of cheaper goods across large distances was not profitable for caravan operators. With productive developments in iron and bronze technologies, newer trade routes-dispensing innovations of civilizations-began to rise slowly.

Maritime Trade

Visible maritime trade between civilizations can be traced back to at least two millennia. Navigation was known in Sumer between the 4th and the 3rd millennium BCE, and was probably known by the Indians and the Chinese people before the Sumerians. The Egyptians had trade routes through the Red sea, importing spices from the "Land of Punt" (East Africa) and from Arabia.

Maritime trade began with safer coastal trade and evolved with the manipulation of the monsoon winds, soon resulting in trade crossing boundaries such as the Arabian Sea and the Bay of Bengal. South Asia had multiple maritime trade routes which connected it to Southeast Asia, thereby making the control of one route resulting in maritime monopoly difficult. Indian connections to various Southeast Asian states buffered it from blockages on other routes. By making use of the maritime trade routes, bulk commodity trade became possible for the Romans in the 2nd century BCE. A Roman trading vessel could span the Mediterranean in a month at one-sixtieth the cost of over-land routes.

Visible Trade Routes

The peninsula of Anatolia lay on the commercial land routes to Europe from Asia as well as the sea route from the Mediterranean to the Black Sea. Records from the 19th century BCE attest to the existence of an Assyrian merchant colony at Kanesh in Cappadocia (now in modern Turkey). Trading networks of the Old World included the Grand Trunk Road of India and the Incense Road of Arabia.

A transportation network consisting of hard-surfaced highways, using concrete made from volcanic ash and lime, was built by the Romans as early as 312 BCE, during the times of the Censor Appius

Claudius Caecus. Parts of the Mediterranean world, Roman Britain, Tigris-Euphrates river system and North Africa fell under the reach of this network at some point of their history.

According to Robert Allen Denemark (2000):

> *"The spread of urban trading networks, and their extension along the Persian Gulf and eastern Mediterranean, created a complex molecular structure of regional foci so that as well as the zonation of core and periphery (originally created around Mesopotamia) there was a series of interacting civilizations: Mesopotamia, Egypt, the Indus Valley; then also Syria, central Anatolia (Hittites) and the Aegean (Minoans and Mycenaeans). Beyond this was a margin which included not only temperate areas such as Europe, but the dry steppe corridor of central Asia. This was truly a world system, even though it occupied only a restricted portion of the western Old World. Whilst each civilization emphasized its ideological autonomy, all were identifiably part of a common world of interacting components."*

These routes-spreading religion, trade and technology-have historically been vital to the growth of urban civilization. The extent of development of cities, and the level of their integration into a larger world system, has often been attributed to their position in various active transport networks.

Historic Trade Routes

Incense Route

The Incense Route served as a channel for trading of Indian, Arabian and East Asian goods. The incense trade flourished from South Arabia to the Mediterranean between roughly the 3rd century BCE to the 2nd century CE. This trade was crucial to the economy of Yemen and the frankincense and myrrh trees were seen as a source of wealth by the its rulers.

Ptolemy II Philadelphus, emperor of Ptolemaic Egypt, may have forged an alliance with the Lihyanites in order to secure the incense route at Dedan, thereby rerouting the incense trade from Dedan to the coast along the Red Sea to Egypt. I.E. S. Edwards connects the Syro-Ephraimite War to the desire of the Israelites and the Aramaeans to control the northern end of the Incense route, which ran up from Southern Arabia and could be tapped by commanding Transjordan.

Gerrha-inhabited by Chaldean exiles from Babylon-controlled the Incense trade routes across Arabia to the Mediterranean and exercised control over the trading of aromatics to Babylon in the 1st century BC. The Nabateans exercised control over the routes along the Incense Route, and their hold was challenged-without success-by Antigonus Cyclops, emperor of Syria and Palestine. The Nabatean control over trade further increased and spread in many directions.

The replacement of Greece by the Roman empire as the administrator of the Mediterranean basin led to the resumption of direct trade with the East and the elimination of the taxes extracted previously by the middlemen of the south. According to Milo Kearney (2003) "The South Arabs in protest took to pirate attacks over the Roman ships in the Gulf of Aden. In response, the Romans destroyed Aden and favoured the Western Abyssinian coast of the Red Sea." Indian ships sailed to Egypt as the maritime routes of Southern Asia were not under the control of a single power.

Pre-Columbian Trade

Some similarities between the Mesoamerican and the Andean cultures suggest that the two regions became a part of a wider world system, as a result of trade, by the 1st millennium BCE. The current academic view is that the flow of goods across the Andean slopes was controlled by institutions distributing locations to local groups, who were then free to access them for trading. This trade across the Andean slopes-described sometimes as "vertical trade"-may have overshadowed the long distance trade between the people of the Andes and the neighboring forests. The Callawaya herbalists traded in tropical plants between 6th and the 10th centuries, while copper was dealt by specialized merchants in the Peruvian valley of Chincha. Long distance trade may have seen local elites resorting to struggle in order for manipulation and control.

Prior to the Inca dominance, specialized long distance merchants provided the highlanders with goods such as gold nuggets, copper hatches, cocoa, salt etc. for redistribution among the locals, and were key players in the politics of the region. Hatchet shaped copper currency was produced by the Peruvian people, in order to obtain valuables from pre Columbian Ecuador. A maritime exchange system stretched from the west coast of Mexico to southernmost Peru, trading mostly in Spondylus, which represented rain, fertility and was considered the principal food of the gods by the people of the Inca empire. Spondylus was used in elite rituals and the effective redistribution of it had political effect in the Andes during the pre-Hispanic times.

Predominantly Overland Routes

Silk Route

The Silk road was one of the first trade routes to join the Eastern and the Western worlds. According to Vadime Elisseeff (2000):

> *"Along the Silk Roads, technology travelled, ideas were exchanged, and friendship and understanding between East and West were experienced for the first time on a large scale. Easterners were exposed to Western ideas and life-styles, and Westerners too, learned about Eastern culture and its spirituality-oriented cosmology. Buddhism as an Eastern religion received international attention through the Silk Roads."*

Cultural interactions patronized often by powerful emperors, such as Kanishka, led to development of art due to introduction of a rich variety of influences. Buddhist missions thrived along the Silk Roads, partly due to the conducive intermixing of trade and cultural values, which created a series of safe stoppages for both the pilgrims and the traders. Among the frequented routes of the Silk Route was the Burmese route extending from Bhamo, which served as a path for Marco Polo's visit to Yunnan and Indian Buddhist missions to Canton in order to establish Buddhist monasteries. This route-often under the presence of hostile tribes-also finds mention in the works of Rashid-al-Din Hamadani.

Grand Trunk Road

The Grand Trunk Road-connecting Calcutta in India to Peshawar in Pakistan-has existed for over two and a half millennia. One of the important trade routes of the world, this road has been a strategic artery with fortresses, halting posts, wells, post offices, milestones and other facilities. Part of this road through Pakistan also coincided with the Silk Road. This highway has been associated with emperors Chandragupta Maurya and Sher Shah Suri, the latter became synonymous with this route due to his role in ensuring the safety of the travellers and the upkeep of the road. Emperor Sher Shah widened and realigned the road to other routes, and provided approximately 1700 roadside inns through his empire. These inns provided free food and lodgings to the travellers regardless of their status.

The British occupation of this road was of special significance for the British Raj in India. Bridges, pathways and newer inns were constructed by the British for the first thirty seven years of their reign since the occupation of Punjab in 1849. The British followed roughly the same alignment as the old routes, and at some places the newer routes ran parallel to the older routes.

Vadime Elisseeff (2000) comments on the Grand Trunk Road:

> *"Along this road marched not only the mighty armies of conquerors, but also the caravans of traders, scholars, artists, and common folk. Together with people, moved ideas, languages, customs, and cultures, not just in one, but in both directions. At different meeting places-permanent as well as temporary-people of different origins and from different cultural backgrounds, professing different faiths and creeds, eating different foods, wearing different clothes, and speaking different languages and dialects would meet one another peacefully. They would understand one another's food, dress, manner, and etiquette, and even borrow words, phrases, idioms and, at times, whole languages from others."*

Amber Road

The Amber Road was a European trade route associated with the trade and transport of amber. Amber satisfied the criteria for long distance trade as it was light in weight and was in high demand for ornamental purposes around the Mediterranean. Before the establishment of Roman control over areas such as Pannonia, the Amber Road was virtually the only route available for long distance trade.

Towns along the Amber Road began to rise steadily during the 1st century CE, despite the troop movements under Titus Flavius Vespasianus and his son Titus Flavius Domitianus. Under the reign of Tiberius Caesar Augustus, the Amber Road was straightened and paved according to the prevailing urban standards. Roman towns began to appear along the road, initially founded near the site of Celtic *oppida*.

The 3rd century saw the Danube river become the principal artery of trade, eclipsing the Amber Road and other commercial routes. The redirection of investment to the Danubian forts saw the towns along the Amber Road growing slowly, though yet retaining their prosperity. The prolonged struggle between the Romans and the barbarians further left its mark on the towns along the Amber Road.

Via Maris

Via Maris, literally Latin for "the way of the sea," was an ancient highway used by the Romans and the Crusaders. The states controlling the Via Maris were in a position to grant access for trade to their own citizens and collect tolls from the outsiders to maintain the trade route. The name *Via Maris* is a Latin translation of a Hebrew phrase related to Isaiah. Due to the Biblical significance of this ancient route, many attempts to find its present day location have been made by Christian pilgrims. 13th century traveller and pilgrim Burchard of Mount Zion refers to the Via Maris route as a way leading along the shore of the Sea of Galilee.

Trans Saharan Trade

Early Muslim writings confirm that the people of West Africa operated a sophisticated network of trade, usually under the authority of a monarch who levied taxes and provided bureaucratic and military support to his kingdom. Sophisticated mechanisms for the economic and political development of the involved African areas were in place before Islam further strengthened trade, towns and government in western Africa. The capital, court and trade of the region find mention in the works of scholar Abu 'Ubayd 'Abd Allah al-Bakri; the mainstay of the trans Saharan trade was gold and salt.

The powerful Saharan tribes, Berber in origin and later adapting to Muslim and Arab cultures, controlled the channels to western Africa by making efficient use of horse-drawn vehicles and pack animals. The Songhai engaged in a struggle against the Saudi dynasty of Morocco over the control of the trans Saharan trade, resulting in damage on both sides and a weak Moroccan victory, further strengthening the uninvolved Saharan tribes. Struggles and disturbances continued till the 14th century, by which the Mande merchants were trading with the Hausa, between Lake Chad and the Niger. Newer trade routes developed following extension of trade.

Predominantly Maritime Routes

Roman-India Routes

The Ptolemaic dynasty had initiated Greco-Roman maritime trade contact with India using the Red Sea ports. The Roman historian Strabo mentions a vast increase in trade following the Roman annexation of Egypt, indicating that monsoon was known and manipulated for trade in his time. By the time of Augustus up to 120 ships were setting sail every year from Myos Hormos to India, trading in a diverse variety of goods. Arsinoe, Berenice Troglodytica and Myos Hormos were the principal Roman ports involved in this maritime trading network, while the Indian ports included Barbaricum, Barygaza, Muziris and Arikamedu.

The Indians were present in Alexandria and the Christian and Jewish settlers from Rome continued to live in India long after the fall of the Roman empire, which resulted in Rome's loss of the Red Sea ports, previously used to secure trade with India by the Greco-Roman world since the time of the Ptolemaic dynasty.

Hanseatic Trade

Shortly before the 12th century the Germans played a relatively modest role in the north European trade. However, this was to change with the development of Hanseatic trade, as a result of which German traders became prominent in the Baltic and the North Sea regions. Following the death of Eric VI of Denmark, German forces attacked and sacked Denmark, bringing with them artisans and merchants under the new administration which controlled the Hansa regions. During the third quarter of the 14th century the Hanseatic trade faced two major difficulties: economic conflict with the Flanders and hostilities with Denmark. These events led to the formation of an organized association of Hanseatic towns, which replaced the earlier union of German merchants. This new Hansa of the towns-aimed at protecting interests of the merchants and trade-became prominent for the next hundred and fifty years.

Philippe Dollinger associates the downfall of the Hansa to a new alliance between Lubeck, Hamburg and Bremen, which outshadowed the older institution. He further sets the date of dissolution of the Hansa at 1630 and concludes that the Hansa was almost entirely forgotten by the end of the 18th century. Scholar Georg Friedrich Sartorius published the first monograph regarding the community in the early years of the 19th century.

Maritime Republics' Mediterranean Trade

The economic growth of Europe around the year 1000, together with the lack of safety on the mainland trading routes, eased the development of major commercial routes along the coast of the Mediterranean. The growing independence of some coastal cities gave them a leading role in this comerce: Maritime Republics (Italian "Repubbliche Marinare") of Venice, Genoa, Amalfi, Pisa and Republic of Ragusa developed their own "empires" in the Mediterranean shores. From the 8th until the 15th century, they held the monopoly of European trade with the Middle East.

The silk and spice trade, involving spices, incense, herbs, drugs and opium, made these Mediterranean city-states phenomenally rich. Spices were among the most expensive and demanded products of the

Middle Ages. They were all imported from Asia and Africa. Muslim traders—mainly descendants of Arab sailors from Yemen and Oman—dominated maritime routes throughout the Indian Ocean, tapping source regions in the Far East and shipping for trading emporiums in India, westward to Ormus in Persian Gulf and Jeddah in the Red Sea. From there, overland routes led to the Mediterranean coasts. Venetian merchants distributed then the goods through Europe until the rise of the Ottoman Empire, that eventually led to the fall of Constantinople in 1453, barring Europeans from important combined-land-sea routes.

Spice Route

As trade between India and the Greco-Roman world increased spices became the main import from India to the Western world, bypassing silk and other commodities. The Indian commercial connection with South East Asia proved vital to the merchants of Arabia and Persia during the 7th century and the 8th century.

The Abbasids used Alexandria, Damietta, Aden and Siraf as entry ports to India and China. Merchants arriving from India in the port city of Aden paid tribute in form of musk, camphor, ambergris and sandalwood to Ibn Ziyad, the sultan of Yemen. Moluccan products shipped across the ports of Arabia to the Near East passed through the ports of India and Sri Lanka. Indian exports of spices find mention in the works of Ibn Khurdadhbeh (850), al-Ghafiqi (1150 AD), Ishak bin Imaran (907) and Al Kalkashandi (14th century). After reaching either the Indian or the Sri Lankan ports, spices were sometimes shipped to East Africa, where they were used for many purposes, including burial rites.

On the orders of Manuel I of Portugal, four vessels under the command of navigator Vasco da Gama rounded the Cape of Good Hope, continuing to the eastern coast of Africa to Malindi to sail across the Indian Ocean to Calicut. The wealth of the Indies was now open for the Europeans to explore; the Portuguese Empire was one of the early European empires to grow from spice trade.

Modern Routes

The modern times saw development of newer means of transport and often controversial free trade agreements, which altered the political and logistical approach prevalent during the Middle Ages. Newer means of transport led to the establishment of new routes, and countries opened up borders to allow trade in mutually agreed goods as per the prevailing free trade agreement. Some old trading route were reopened during the modern times, although in different political and logistical scenarios. The entry of harmful foreign pollutants by the way of trade routes has been a cause of alarm during the modern times. A conservative estimate stresses that future damages from harmful animal and plant diseases may be as high as 134 billion US dollars in the absence of effective measures to prevent the introduction of unwanted pests through various trade routes.

Wagonway Routes

Networks, like the Santa Fe Trail and the Oregon Trail, became prominent in the United States with wagon trains gaining popularity as a mode of long distance overland transportation for both people and goods. The Oregon-California routes were highly organized with planned rendezvous locations and essential supplies. The settlers in the United States used these wagon trains-sometimes made up of 100 of more Conestoga wagons-for westward emigration during the 18th and the 19th centuries. Among the challenges faced by the wagon route operators were crossing rivers, mountains and hostile Native Americans. Preparations were also made according to the weather and protection of trade and travellers was ensured by a few guards on horseback. Wagon freighting was also essential to American growth until it was replaced by the railroad and the truck.

Railway Routes

The 1844 Railway act of England compelled at least one train to a station every day with the third class fares priced at a penny a mile. Trade benefited as the workers and the lower classes had the ability to travel to other towns frequently. Suburban communities began to develop and towns began to spread outwards. The British constructed a vast railway network in India, but it was considered to serve a strategic purpose in addition to the commercial purpose. The efficient use of rail routes helped in the unification of the United States of America.

The modern times saw nations struggle for the control of rail routes: The Trans-Siberian Railway was intended to be used by the Russian government for control of Manchuria and later China; the German forces wanted to establish Berlin-Baghdad Railway in order to influence the Near East; and the Austrian government planned a route from Vienna to Salonika for control of the Balkans. According to the Encyclopaedia Britannica (2002):

Railroads reached their maturity in the early 20th century, as trains carried the bulk of land freight and passenger traffic in the industrialized countries of the world. By the mid-20th century, however, they had lost their preeminent position. The private automobile had replaced the railroad for short passenger trips, while the airplane had usurped it for long-distance travel, especially in the United States.

Railroads remained effective, however, for transporting people in high-volume situations, such as commuting between the centres of large cities and their suburbs, and medium-distance travel of less than about 300 miles between urban centres. Although railroads have lost much of the general-freight-carrying business to semi-trailer trucks, they remain the best means of transporting large volumes of such bulk commodities as coal, grain, chemicals, and ore over long distances. The development of containerization has made the railroads more effective in handling finished merchandise at relatively high speeds. In addition, the introduction of piggyback flatcars, in which truck trailers are transported long distances on specially-designed cars, has allowed railroads to regain some of the business lost to trucking.

Modern Road Networks

The advent motor vehicles created a demand for better use of highways. Roads evolved into two way roads, expressways, freeways and tollways during the modern times. Existing roads were developed and highways were designed according to intended use.

Trucks came into widespread use in the Western World during World War I, and quickly gained reputation as a means of long distance transportation of goods. Modern highways, such as the Trans-Canada Highway, Highway 1 (Australia) and Pan-American Highway allowed transport of goods and services across great distances. Automobiles continue to play a crucial role in the economies of the Industrialized countries, resulting in rise of businesses such as motor freight operation and truck transportation. The emission rate for cars using highways has been on a decline between 1975 and 1995 due to regulations and the introduction of unleaded petrol. This trend is especially notable since there has been a growth in vehicles and vehicle miles travelled by automobiles using these highways.

Modern Maritime Routes

A consistent shift from land based trade to sea based trade has been recorded since the last three millennia. The strategic advantages of port cities as trading centres are many: they are both less dependent on vital connections and less vulnerable to blockages. Oceanic ports can help forge trading relationships with other parts of the world easily.

Modern maritime trade routes-sometimes in the form of artificial canals like the Suez Canal-had visible impact on the economic and political standing of nations. The opening of the Suez Canal altered British interactions with the colonies of the British Empire as the dynamics of transportation, trade and communication had now changed drastically. Other waterways, like the Panama Canal played an important role in the histories of many nations. Inland water transportation remained significantly important even as the advent of railroads and automobiles resulted in a steady decline of canals. Inland water transport is still used for the transportation of bulk commodities e.g. grains, coal, and ore.

Waterway commerce was historically important to Europe, particularly to Russia. According to the *Encyclopaedia Britannica* (2002): "Russia has been a significant beneficiary. Not only have inland waterways opened vast areas of its interior to development, but Moscow-linked to the White, Baltic, Black, Caspian, and Azov seas by canals and rivers-has become a major inland port."

Oil spills are recorded both in case of maritime routes and pipeline routes to the main refineries. Oil spills, amounting to as much as 7. 56 billion liters of oil entering the oceans every year, occur due to damaged equipment or human error.

Free Trade Areas

Historically, governments followed a policy of protection of trade. International Free Trade became visible in 1860 with the Anglo-French commercial treaty and the sentiment further gained momentum during the post World War II era. According to The Columbia Encyclopedia, Sixth Edition:

> *"After World War II, strong sentiment developed throughout the world against protection and high tariffs and in favour of freer trade. The results were new organizations and agreements on international trade such as the General Agreement on Tariffs and Trade (1948), the Benelux Economic Union (1948), the European Economic Community (Common Market, 1957), the European Free Trade Association (1959), Mercosur (the Southern Cone Common Market, 1991), and the World Trade Organization (1995). In 1993, the North American Free Trade Agreement (NAFTA) was approved by the governments of Canada, Mexico, and the United States. In the early 1990s, the nations of the European Union (the successor organization to the Common Market) undertook to remove all barriers to the free movement of trade and employment across their mutual borders."*

On May 2004, the United States of America signed the American Free trade Agreement with five Central American nations.

Air Routes

Air transport has become an indispensable part of the modern society. People having been using air transport both for long and middle distances, with the average route length of long distances being 720 kilometers in Europe and 1220 kilometers in the US. This enormous industry annually carries 1600 million passengers worldwide, and covers a 15 million kilometre network with an annual turnover of 260 billion dollars.

The national, international and global economies are linked to this mode of transportation, making it vital to many other industries. Newer trends of liberalization of trade have further led to establishment of routes among nations bound by agreements. One such example is the American Open Skies policy, which led to greater openness in many international markets, but some international restrictions have survived even during the present times.

Express delivery through international cargo airlines touched US $ 20 billion in 1998 and, according to the World Trade Organization, is expected to triple in 2015. In 1998, 50 pure cargo service companies operated internationally. Civil Aviation

Civil aviation is one of two major categories of flying, representing all non-military aviation, both private and commercial. Most of the countries in the world are members of the International Civil Aviation Organization (ICAO) and work together to establish common standards and recommended practices for civil aviation through that agency.

Civil aviation includes two major categories:

- Scheduled air transport, including all passenger and cargo flights operating on regularly-scheduled routes; and
- General aviation (GA), including all other civil flights, private or commercial.

Although scheduled air transport is the larger operation in terms of passenger numbers, GA is larger in the number of flights (and flight hours, in the U. S.) In the U. S., GA carries 166 million passengers each year, more than any individual airline, though less than all the airlines combined. Some countries also make a regulatory distinction based on whether aircraft are flown for hire:

- Commercial aviation includes most or all flying done for hire, particularly scheduled service on airlines; and

- Private aviation includes pilots flying for their own purposes (recreation, business meetings, etc.) without receiving any kind of remuneration.

All scheduled air transport is commercial, but general aviation can be either commercial or private. Normally, the pilot, aircraft, and operator must all be authorized to perform commercial operations through separate commercial licensing, registration, and operation certificates.

Civil Aviation Authorities

The Convention on International Civil Aviation (the *Chicago Convention*) was originally established in 1944: it states that signatories should collectively work to harmonize and standardize the use of airspace for safety, efficiency and regularity of air transport. All the States signatory to the Chicago Convention, now 188, are obliged to implement the Standards and Recommended Practices (SARPs) of the Convention.

Each signatory country has a Civil Aviation Authority (CAA) (such as the FAA in the United States) to oversee the following areas of civil aviation:

- Personnel Licensing — regulating the basic training and issuance of licenses and certificates.
- Flight Operations — carrying out safety oversight of commercial operators.
- Airworthiness — issuing certificates of registration and certificates of airworthiness to civil aircraft, and overseeing the safety of maintenance organizations.
- Aerodromes — designing and constructing aerodrome facilities.
- Air Traffic Services — managing the traffic inside of a country's airspace.

Airline

An airline provides air transport services for passengers or freight, generally with a recognized operating certificate or license. Airlines lease or own their aircraft with which to supply these services and may form partnerships or alliances with other airlines for mutual benefit.

Airlines vary from those with a single aircraft carrying mail or cargo, through full-service international airlines operating hundreds of aircraft. Airline services can be categorized as being intercontinental, intra continental, domestic, or international and may be operated as scheduled services or charters.

History

The First Airlines

DELAG, *Deutsche Luftschiffahrts-Aktiengesellschaft* was the world's first airline. It was founded on November 16, 1909 with government assistance, and operated airships manufactured by The Zeppelin Corporation. Its headquarters were in Frankfurt. (Note: Americans, such as Rufus Porter and Frederick Marriott, attempted to start airlines using airships in the mid-19th century, focusing on the New York–California route. Those attempts floundered due to such mishaps as the aircraft catching fire and the aircraft being ripped apart by spectators.) The five oldest non-dirigible airlines that still exist are Netherlands' KLM, Colombia's Avianca, Australia's Qantas, Czech Republic's Czech Airlines, and Mexico's Mexicana. KLM first flew in May 1920 while Qantas (for the Queensland and Northern Territory Aerial Services Limited) was founded in Queensland, Australia, in late 1920.

U. S. Airline Industry

Early Development

Tony Jannus conducted the United State's first scheduled commercial airline flight on 1 January 1914 for the St. Petersburg-Tampa Airboat Line. Thc 23-minute flight travelled between St. Petersburg, Florida and Tampa, Florida, passing some 50 feet above Tampa Bay in Jannus' Benoist XIV biplane flying boat. Chalk's Airlines (now Chalk's International Airlines) began service between Miami and Bimini in the Bahamas in February 1919. Now based in Ft. Lauderdale, Chalk's claims to be the oldest continuously operating airline in the United States.

Following World War I, the United States found itself swamped with aviators. Many decided to take their war-surplus aircraft on barnstorming campaigns, performing acrobatic maneuvers to woo crowds. In 1918, the United States Postal Service won the financial backing of Congress to begin experimenting with air mail service, initially using Curtiss Jenny aircraft that had been procured by the United States Army for reconnaissance missions on the Western Front. Private operators were the first to fly the mail but due to numerous accidents the US Army was tasked with mail delivery. During the course of the Army's involvement they proved to be too unreliable and lost their air mail duties. By the mid-1920s, the Postal Service had developed its own air mail network, based on a transcontinental backbone between New York and San Francisco. To supplant this service, they offered twelve contracts for spur routes to independent bidders. Some of the carriers that won these routes would, through time and mergers, evolve into Delta Air Lines, Braniff Airways, American Airlines, United Airlines (originally a division of Boeing), Trans World Airlines, Northwest Airlines, and Eastern Air Lines.

Service during the early 1920s was sporadic: most airlines at the time were focused on carrying bags of mail. In 1925, however, the Ford Motor Company bought out the Stout Aircraft Company and began construction of the all-metal Ford Trimotor, which became the first successful American airliner. With a 12-passenger capacity, the Trimotor made passenger service potentially profitable. Air service was seen as a supplement to rail service in the American transportation network.

At the same time, Juan Trippe began a crusade to create an air network that would link America to the world, and he achieved this goal through his airline, Pan American World Airways, with a fleet of flying boats that linked Los Angeles to Shanghai and Boston to London. Pan Am and Northwest Airways (which began flights to Canada in the 1920s) were the only U. S. airlines to go international before the 1940s.

With the introduction of the Boeing 247 and Douglas DC-3 in the 1930s, the U. S. airline industry was generally profitable, even during the Great Depression. This trend continued until the beginning of World War II.

Development Since 1945

As governments met to set the standards and scope for an emergent civil air industry toward the end of the war, the U. S. took a position of maximum operating freedom; U. S. airline companies were not as hard-hit as European and the few Asian ones had been. This preference for "open skies" operating regimes continues, within limitations, to this day.

World War II, like World War I, brought new life to the airline industry. Many airlines in the Allied countries were flush from lease contracts to the military, and foresaw a future explosive demand for civil air transport, for both passengers and cargo. They were eager to invest in the newly emerging flagships of air travel such as the Boeing Stratocruiser, Lockheed Constellation, and Douglas DC-6. Most of these new aircraft were based on American bombers such as the B-29, which had spearheaded research into new technologies such as pressurization. Most offered increased efficiency from both added speed and greater payload.

In the 1950s, the De Havilland Comet, Boeing 707, Douglas DC-8, and Sud Aviation Caravelle became the first flagships of the Jet Age in the West, while the Soviet Union bloc had Tupolev Tu-104 and Tupolev Tu-124 in the fleets of state-owned carriers such as Czechoslovak ESA, Soviet Aeroflot and East-German Interflug. The Vickers Viscount and Lockheed L-188 Electra inaugurated turboprop transport.

The next big boost for the airlines would come in the 1970s, when the Boeing 747, McDonnell Douglas DC-10, and Lockheed L-1011 inaugurated widebody ("jumbo jet") service, which is still the standard in international travel. The Tupolev Tu-144 and its Western counterpart, Concorde, made supersonic travel a reality. Concorde first flew in 1969 and operated through 2003. In 1972, Airbus began producing Europe's most commercially successful line of airliners to date. The added efficiencies for these aircraft were often not in speed, but in passenger capacity, payload, and range. Airbus also features modern electronic cockpits that were common across their aircraft to enable pilots to fly multiple models with minimal cross-training.

1978's U. S. airline industry deregulation lowered barriers for new airlines just as a downturn occurred. New start-ups entered during the downturn, during which time they found aircraft and funding, contracted hangar and maintenance services, trained new employees, and recruited laid off staff from other airlines. As the business cycle returned to normalcy, major airlines dominated their routes through aggressive pricing and additional capacity offerings, often swamping new startups. Only America West Airlines (which has since merged with US Airways) remained a significant survivor from this new entrant era, as dozens, even hundreds, have gone under.

In many ways, the biggest winner in the deregulated environment was the air passenger. Indeed, the U. S. witnessed an explosive growth in demand for air travel, as many millions who had never or rarely flown before became regular fliers, even joining frequent flyer loyalty programs and receiving free flights and other benefits from their flying. New services and higher frequencies meant that business fliers could fly to another city, do business, and return the same day, for almost any point in the country. Air travel's advantages put intercity bus lines under pressure, and most have withered away.

By the 1980s, almost half of the total flying in the world took place in the U. S., and today the domestic industry operates over 10,000 daily departures nationwide.

Toward the end of the century, a new style of low cost airline emerged, offering a no-frills product at a lower price. Southwest

Airlines, Jet Blue, Air Tran Airways, Skybus Airlines and other low-cost carriers began to represent a serious challenge to the so-called "legacy airlines", as did their low-cost counterparts in many other countries. Their commercial viability represented a serious competitive threat to the legacy carriers. However, of these, ATA and Skybus have since ceased operations. Thus the last 50 years of the airline industry have varied from reasonably profitable, to devastatingly depressed. As the first major market to deregulate the industry in 1978, U. S. airlines have experienced more turbulence than almost any other country or region. Today, American Airlines is the only U. S. legacy carrier to survive bankruptcy-free.

The Airline "Bailout"

Congress passed the Air Transportation Safety and System Stabilization Act (P. L. 107-42) in response to a severe liquidity crisis facing the industry in the aftermath of the September 11th terrorist attacks. Congress sought to compensate carriers for both the cost of the four-day federal shutdown of the airlines and the incremental losses incurred through December 31, 2001 as a result of the terrorist attacks. Congress expressly sought to preserve a viable, safe, and efficient air transportation system.

In recognition of the essential national economic role of a healthy aviation system, Congress authorized partial compensation of up to $5 billion in cash subject to review by the Department of Transportation and up to $10 billion in loan guarantees subject to review by a newly created Air Transportation Stabilization Board (ATSB). The applications to DOT for reimbursements were subjected to rigorous multi-year reviews not only by DOT program personnel but also by the Government Accountability Office and the DOT Inspector General.

Ultimately, the federal government provided $4. 6 billion in one-time, subject-to-income-tax cash reimbursements to 427 U. S. air carriers, including numerous charter and cargo carriers. (Passenger carriers operating scheduled service received approximately $4 billion, subject to tax.) In addition, the ATSB approved loan guarantees to six airlines totaling approximately $1. 6 billion. Data from the Treasury Department show that taxpayers eventually recouped the $1. 6 billion and a profit of $339 million from the fees, interest and stock associated with loan guarantees.

European Airline Industry

The first countries in Europe to embrace air transport were Finland, France, Germany, the Netherlands and the United Kingdom.

KLM, the oldest carrier still operating under its original name, was founded in 1919. The first flight (operated on behalf of KLM by Aircraft Transport and Travel) transported two English passengers to Schiphol, Amsterdam from London in 1920. Like other major European airlines of the time, KLM's early growth depended heavily on the needs to service links with far-flung colonial possessions (Dutch Indies). It is only after the loss of the Dutch Empire that KLM found itself based at a small country with few potential passengers, depending heavily on transfer traffic, and was one of the first to introduce the hub-system to facilitate easy connections. France began an air mail service to Morocco in 1919 that was bought out in 1927, renamed Aeropostale, and injected with capital to become a major international carrier. In 1933, Aeropostale went bankrupt, was nationalized and merged with several other airlines into what became Air France.

In Finland, the charter establishing Aero O/Y (now Finnair, one of the oldest still-operating airlines in the world) was signed in the city of Helsinki on September 12, 1923. Junkers F 13 D-335 became the first aircraft of the company, when Aero took delivery of it on March 14, 1924. The first flight was between Helsinki and Tallinn, capital of Estonia, and it took place on March 20, 1924, one week later.

Germany's Lufthansa began in 1926. Lufthansa, unlike most other airlines at the time, became a major investor in airlines outside of Europe, providing capital to Varig and Avianca. German airliners built by Junkers, Dornier, and Fokker were the most advanced in the world at the time. In 1931, the airship Graf Zeppelin began offering regular scheduled passenger service between Germany and South America, usually every two weeks, which continued until 1937. In 1936, the airship Hindenburg entered passenger service and successfully crossed the Atlantic 36 times before crashing at Lakehurst, New Jersey on May 6, 1937.

The British company Aircraft Transport and Travel commenced a London to Paris service on August 25, 1919, this was the world's first regular international flight. The United Kingdom's flag carrier during this period was Imperial Airways, which became BOAC (British Overseas Airways Co.) in 1939. Imperial Airways used huge Handley-Page biplanes for routes between London, the Middle East, and India: images of Imperial aircraft in the middle of the Rub'al Khali, being maintained by Bedouins, are among the most famous pictures from the heyday of the British Empire.

In Soviet Union the Chief Administration of the Civil Air Fleet was established in 1921. One of its first acts was to help found Deutsch-

Russische Luftverkehrs A. G. (Deruluft), a German-Russian joint venture to provide air transport from Russia to the West. Domestic air service began around the same time, when Dobrolyot started operations on 15 July 1923 between Moscow and Nizhni Novgorod. Since 1932 all operations had been carried under the name Aeroflot. By the end of the 1930s Aeroflot had become the world's largest airline, employing more than 4,000 pilots and 60,000 other service personnel and operating around 3,000 aircraft (of which 75% were considered obsolete by its own standards). During the Soviet era Aeroflot was synonymous with Russian civil aviation, as it was the only air carrier. It became the first airline in the world to operate sustained regular jet services on 15 September 1956 with the Tupolev Tu-104.

Deregulation

Deregulation of the European Union airspace in the early 1990s has had substantial effect on structure of the industry there. The shift towards 'budget' airlines on shorter routes has been significant. Airlines such as Easy Jet and Ryanair have grown at the expense of the traditional national airlines.

There has also been a trend for these national airlines themselves to be privatised such as has occurred for Aer Lingus and British Airways. Other national airlines, including Italy's Alitalia, have suffered-particularly with the rapid increase of oil prices in early 2008.

Asian Airline Industry

Although Philippine Airlines (PAL) was officially founded on February 26, 1941, its license to operate as an airliner was derived from merged Philippine Aerial Taxi Company (PATCO) established by mining magnate Emmanuel N. Bachrach in December 3, 1930, making it as Asia's oldest scheduled carrier still in operation. Commercial air service commenced three weeks later from Manila to Baguio, making it Asia's first airline route. Bachrach's death in 1937 paved the way for its eventual merger with Philippine Airlines in March 1941 and made it Asia's oldest airline. It is also the oldest airline in Asia still operating under its current name. Bachrach's majority share in PATCO was bought by beer magnate Andres R. Soriano in 1939 upon the advice of General Douglas McArthur and later merged with newly formed Philippine Airlines with PAL as the surviving entity. Soriano has controlling interest in both airlines before the merger. PAL restarted service on March 15, 1941 with a single Beech Model 18 NPC-54 aircraft, which started its daily services between Manila (from Nielson Field) and Baguio, later to expand with larger aircraft such as the DC-3 and Vickers Viscount.

India was also one of the first countries to embrace civil aviation. One of the first West Asian airline companies was Air India, which had its beginning as Tata Airlines in 1932, a division of Tata Sons Ltd. (now Tata Group). The airline was founded by India's leading industrialist, JRD Tata. On October 15, 1932, J. R. D. Tata himself flew a single engined De Havilland Puss Moth carrying air mail (postal mail of Imperial Airways) from Karachi to Bombay via Ahmedabad. The aircraft continued to Madras via Bellary piloted by Royal Air Force pilot Nevill Vincent. Tata Airlines was also one of the world's first major airlines which began its operations without any support from the Government. With the outbreak of World War II, the airline presence in Asia came to a relative halt, with many new flag carriers donating their aircraft for military aid and other uses. Following the end of the war in 1945, regular commercial service was restored in India and Tata Airlines became a public limited company on July 29, 1946 under the name Air India. After the independence of India, 49% of the airline was acquired by the Government of India. In return, the airline was granted status to operate international services from India as the designated flag carrier under the name Air India International.

On July 31, 1946, a chartered Philippine Airlines (PAL) DC-4 ferried 40 American servicemen to Oakland, California from Nielson Airport in Makati City with stops in Guam, Wake Island, Johnston Atoll and Honolulu, Hawaii, making PAL the first Asian airline to cross the Pacific Ocean. A regular service between Manila and San Francisco was started in December. It was during this year that the airline was designated as the flag carrier of Philippines.

During the era of decolonization, newly-born Asian countries started to embrace air transport. Among the first Asian carriers during the era were Cathay Pacific of Hong Kong (founded in September 1946), Orient Airways (later Pakistan International Airlines; founded in October 1946), Malayan Airlines (later Singapore and Malaysia Airlines; founded in 1947), Garuda Indonesia in 1949, Japan Airlines in 1951, and Korean Air in 1962.

Latin American Airline Industry

Among the first countries to have regular airlines in Latin America were Colombia with Avianca, Brazil with Varig, Chile with LAN Chile (today LAN Airlines), Dominican Republic with Air Dominicana, Mexico with Mexicana de Aviacion, and TACA as a brand of several airlines of Central American countries (Honduras, El Salvador, Costa Rica, Guatemala and Nicaragua). All the previous airlines started regular operations before World War II.

The air travel market has evolved rapidly over recent years in Latin America. Some industry estimations over 2000 new aircraft will begin service over the next five years in this region. These airlines serve domestic flights within their countries, as well as connections within Latin America and also overseas flights to North America, Europe, Australia, Africa and Asia.

Just three airlines: LAN (Latin American Networks), Oceanair and TAM Airlines have international subsidiaries with Chile as the central operation along with Peru, Ecuador, Argentina and some operations in the Dominican Republic and TAM with TAM Mercosur have a base in Asuncion, Paraguay. Avianca have the control of Oceanair, VIP Airlines and also have an estrategic alliance with TACA.

The three main hubs in Latin America are Mexico City in Mexico, Gaurulhos International Airport in Brazil and Bogota.

Regulatory Considerations

National

Many countries have national airlines that the government owns and operates. Fully private airlines are subject to a great deal of government regulation for economic, political, and safety concerns. For instance, governments often intervene to halt airline labour actions in order to protect the free flow of people, communications, and goods between different regions without compromising safety. The United States, Australia, and to a lesser extent Brazil, Mexico, the United Kingdom and Japan have "deregulated" their airlines. In the past, these governments dictated airfares, route networks, and other operational requirements for each airline. Since deregulation, airlines have been largely free to negotiate their own operating arrangements with different airports, enter and exit routes easily, and to levy airfares and supply flights according to market demand.

The entry barriers for new airlines are lower in a deregulated market, and so the U. S. has seen hundreds of airlines start up (sometimes for only a brief operating period). This has produced far greater competition than before deregulation in most markets, and average fares tend to drop 20% or more.

The added competition, together with pricing freedom, means that new entrants often take market share with highly reduced rates that, to a limited degree, full service airlines must match. This is a major constraint on profitability for established carriers, which tend to have a higher cost base. As a result, profitability in a deregulated market is uneven for most airlines. These forces have caused some major airlines to go out of business, in addition to most of the poorly established new entrants.

International

Groups such as the International Civil Aviation Organization establish worldwide standards for safety and other vital concerns. Most international air traffic is regulated by bilateral agreements between countries, which designate specific carriers to operate on specific routes. The model of such an agreement was the Bermuda Agreement between the US and UK following World War II, which designated airports to be used for transatlantic flights and gave each government the authority to nominate carriers to operate routes.

Bilateral agreements are based on the "freedoms of the air", a group of generalized traffic rights ranging from the freedom to overfly a country to the freedom to provide domestic flights within a country (a very rarely granted right known as cabotage). Most agreements permit airlines to fly from their home country to designated airports in the other country: some also extend the freedom to provide continuing service to a third country, or to another destination in the other country while carrying passengers from overseas.

In the 1990s, "open skies" agreements became more common. These agreements take many of these regulatory powers from state governments and open up international routes to further competition. Open skies agreements have met some criticism, particularly within the European Union, whose airlines would be at a comparative disadvantage with the United States' because of cabotage restrictions.

Economic Considerations

Historically, air travel has survived largely through state support, whether in the form of equity or subsidies. The airline industry as a whole has made a cumulative loss during its 100-year history, once the costs include subsidies for aircraft development and airport construction.

One argument is that positive externalities, such as higher growth due to global mobility, outweigh the microeconomic losses and justify continuing government intervention. A historically high level of government intervention in the airline industry can be seen as part of a wider political consensus on strategic forms of transport, such as highways and railways, both of which receive public funding in most parts of the world. Profitability is likely to improve in the future as privatization continues and more competitive low-cost carriers proliferate. Although many countries continue to operate state-owned or parastatal airlines, many large airlines today are privately owned and are therefore governed by microeconomic principles in order to maximize shareholder profit.

Ticket Revenue

Airlines assign prices to their services in an attempt to maximize profitability. The pricing of airline tickets has become increasingly complicated over the years and is now largely determined by computerized yield management systems.

Because of the complications in scheduling flights and maintaining profitability, airlines have many loopholes that can be used by the knowledgeable traveller. Many of these airfare secrets are becoming more and more known to the general public, so airlines are forced to make constant adjustments. Most airlines use differentiated pricing, a form of price discrimination, in order to sell air services at varying prices simultaneously to different segments. Factors influencing the price include the days remaining until departure, the booked load factor, the forecast of total demand by price point, competitive pricing in force, and variations by day of week of departure and by time of day. Carriers often accomplish this by dividing each cabin of the aircraft (first, business and economy) into a number of travel classes for pricing purposes.

A complicating factor is that of origin-destination control ("O&D control"). Someone purchasing a ticket from Melbourne to Sydney (as an example) for AU$200 is competing with someone else who wants to fly Melbourne to Los Angeles through Sydney on the same flight, and who is willing to pay AU$1400. Should the airline prefer the $1400 passenger, or the $200 passenger plus a possible Sydney-Los Angeles passenger willing to pay $1300? Airlines have to make hundreds of thousands of similar pricing decisions daily.

The advent of advanced computerized reservations systems in the late 1970s, most notably Sabre, allowed airlines to easily perform cost-benefit analyses on different pricing structures, leading to almost perfect price discrimination in some cases (that is, filling each seat on an aircraft at the highest price that can be charged without driving the consumer elsewhere).

The intense nature of airfare pricing has led to the term "fare war" to describe efforts by airlines to undercut other airlines on competitive routes. Through computers, new airfares can be published quickly and efficiently to the airlines' sales channels. For this purpose the airlines use the Airline Tariff Publishing Company (ATPCO), who distribute latest fares for more than 500 airlines to Computer Reservation Systems across the world.

The extent of these pricing phenomena is strongest in "legacy" carriers. In contrast, low fare carriers usually offer preannounced and simplified price structure, and sometimes quote prices for each leg of a trip separately. Computers also allow airlines to predict, with some accuracy, how many passengers will actually fly after making a reservation to fly. This allows airlines to overbook their flights enough to fill the aircraft while accounting for "no-shows," but not enough (in most cases) to force paying passengers off the aircraft for lack of seats. Since an average of S! of all seats are flown empty, stimulative pricing for low demand flights coupled with overbooking on high demand flights can help reduce this figure. This is especially crucial during tough economic times as airlines undertake massive cuts to ticket prices in order to retain demand.

Operating Costs

Full-service airlines have a high level of fixed and operating costs in order to establish and maintain air services: labour, fuel, airplanes, engines, spares and parts, IT services and networks, airport equipment, airport handling services, sales distribution, catering, training, aviation insurance and other costs. Thus all but a small percentage of the income from ticket sales is paid out to a wide variety of external providers or internal cost centres.

Moreover, the industry is structured so that airlines often act as tax collectors. Airline fuel is untaxed because of a series of treaties existing between countries. Ticket prices include a number of fees, taxes and surcharges beyond the control of airlines. Airlines are also responsible for enforcing government regulations. If airlines carry passengers without proper documentation on an international flight, they are responsible for returning them back to the original country.

Analysis of the 1992–1996 period shows that every player in the air transport chain is far more profitable than the airlines, who collect and pass through fees and revenues to them from ticket sales. While airlines as a whole earned 6% return on capital employed (2-3. 5% less than the cost of capital), airports earned 10%, catering companies 10-13%, handling companies 11-14%, aircraft lessors 15%, aircraft manufacturers 16%, and global distribution companies more than 30%. (Source: Spinetta, 2000, quoted in Doganis, 2002)

In contrast, Southwest Airlines has been the most profitable of airline companies since 1973. The widespread entrance of a new breed of low cost airlines beginning at the turn of the century has accelerated the demand that full service carriers control costs. Many of these low cost companies emulate Southwest Airlines in various respects, and like Southwest, they are able to eke out a consistent profit throughout all phases of the business cycle. As a result, a shakeout of airlines is occurring in the U. S. and elsewhere. United Airlines, Continental Airlines (twice), US Airways (twice), Delta Air Lines, and Northwest Airlines have all declared Chapter 11 bankruptcy. Some argue that it would be far better for the industry as a whole if a wave of actual closures were to reduce the number of "undead" airlines competing with healthy airlines while being artificially protected from creditors via bankruptcy law. On the other hand, some have pointed out that the reduction in capacity would be short lived given that there would be large quantities of relatively new aircraft that bankruptcies would want to get rid of and would re-enter the market either as increased fleets for the survivors or the basis of cheap planes for new startups.

Where an airline has established an engineering base at an airport then there may be considerable economic advantages in using that same airport as a preferred focus (or "hub") for its scheduled flights.

Assets and Financing

Airline financing is quite complex, since airlines are highly leveraged operations. Not only must they purchase (or lease) new airliner bodies and engines regularly, they must make major long-term fleet decisions with the goal of meeting the demands of their markets while producing a fleet that is relatively economical to operate and maintain. Compare Southwest Airlines and their reliance on a single airplane type (the Boeing 737 and derivatives), with the now defunct Eastern Air Lines which operated 17 different aircraft types, each with varying pilot, engine, maintenance, and support needs.

A second financial issue is that of hedging oil and fuel purchases, which are usually second only to labour in its relative cost to the company. However, with the current high fuel prices it has become the largest cost to an airline. While hedging instruments can be expensive, they can easily pay for themselves many times over in periods of increasing fuel costs, such as in the 2000–2005 period. In view of the congestion apparent at many international airports, the ownership of slots at certain airports (the right to take-off or land an aircraft at a particular time of day or night) has become a significant tradable asset for many airlines. Clearly take-off slots at popular times of the day can be critical in attracting the more profitable business traveller to a given airline's flight and in establishing a competitive advantage against a competing airline. If a particular city has two or more airports, market forces will tend to attract the less profitable routes, or those on which competition is weakest, to the less congested airport, where slots are likely to be more available and therefore cheaper. Other factors, such as surface transport facilities and onward connections, will also affect the relative appeal of different airports and some long distance flights may need to operate from the one with the longest runway.

Airline Partnerships

Code sharing is the most common type of airline partnership; it involves one airline selling tickets for another airline's flights under its own airline code. An early example of this was Japan Airlines' code sharing partnership with Aeroflot in the 1960s on flights from Tokyo to Moscow: Aeroflot operated the flights using Aeroflot aircraft, but JAL sold tickets for the flights as if they were JAL flights. This practice allows airlines to expand their operations, at least on paper, into parts of the world where they cannot afford to establish bases or purchase aircraft. Another example was the Austrian-Sabena partnership on the Vienna-Brussels-New York JFK route during the late '60s, using a Sabena Boeing 707 with Austrian colours.

Since airline reservation requests are often made by city-pair (such as "show me flights from Chicago to Düsseldorf"), an airline who is able to code share with another airline for a variety of routes might be able to be listed as indeed offering a Chicago-Dusseldorf flight. The passenger is advised however, that Airline 1 operates the flight from say Chicago to Amsterdam, and Airline 2 operates the continuing flight (on a different airplane, sometimes from another terminal) to Düsseldorf. Thus the primary rationale for code sharing is to expand one's service offerings in city-pair terms so as to increase sales. A more recent development is the airline alliance, which became prevalent in the 1990s. These alliances can act as virtual mergers to get around government restrictions. Groups of airlines such as the Star Alliance, Oneworld, and Sky Team coordinate their passenger service programs (such as lounges and frequent flyer programs), offer special interline tickets, and often engage in extensive codesharing (sometimes systemwide). These are increasingly integrated business combinations—sometimes including cross-equity arrangements—in which products, service standards, schedules, and airport facilities are standardized and combined for higher efficiency. One of the first airlines to start an alliance with another airline was KLM, who partnered with

Northwest Airlines. Both airlines later entered the Sky Team alliance after the fusion of KLM and Air France in 2004.

Often the companies combine IT operations, buy fuel, or purchase airplanes as a bloc in order to achieve higher bargaining power. However, the alliances have been most successful at purchasing invisible supplies and services, such as fuel. Airlines usually prefer to purchase items visible to their passengers to differentiate themselves from local competitors. If an airline's main domestic competitor flies Boeing airliners, then the airline may prefer to use Airbus aircraft regardless of what the rest of the alliance chooses.

Environmental Impacts

Aircraft engines emit noise pollution, gases and particulate emissions, and contribute to global warming and global dimming,.

Modern turbofan and turboprop engines are considerably more fuel-efficient and less polluting than earlier models. However, despite this, the rapid growth of air travel in recent years contributes to an increase in total pollution attributable to aviation, offsetting some of the reductions achieved by automobiles. In the EU greenhouse gas emissions from aviation increased by 87% between 1990 and 2006.

CO2 emissions from the jet fuel burned per passenger on an average 3200 kilometers (1992 miles) airline flight is about 353 kilograms (776 pounds). Loss of natural habitat potential associated with the jet fuel burned per passenger on a 3200 kilometers (1992 miles) airline flight is estimated to be 250 square meters (2700 square feet).

In the context of purported climate change and peak oil, there is a debate about possible taxation of air travel and the inclusion of aviation in an emissions trading scheme, with a view to ensuring that the total external costs of aviation are taken into account. The airline industry is responsible for about 11 percent of greenhouse gases emitted by the U. S. transportation sector. Boeing estimates that biofuels could reduce flight-related greenhouse-gas emissions by 60 to 80 percent. The solution would be blending algae fuels with existing jet fuel:

- Boeing and Air New Zealand are collaborating with leading Brazilian biofuels maker Tecbio and Aquaflow Bionomic of New Zealand and other jet biofuel developers around the world.
- Virgin Atlantic and Virgin Green Fund are looking into the technology as part of a biofuels initiative.
- KLM has made the first commercial flight with bio-fuel in 2009.

Call Signs

Each operator of a scheduled or charter flight uses an airline call sign when communicating with airports or air traffic control centres. Most of these call-signs are derived from the airline's trade name, but for reasons of history, marketing, or the need to reduce ambiguity in spoken English (so that pilots do not mistakenly make navigational decisions based on instructions issued to a different aircraft), some airlines and air forces use call-signs less obviously connected with their trading name. For example, British Airways uses a *Speedbird* call-sign, named after the logo of its predecessor, BOAC, while Sky Europe used *Relax*.

Airline Personnel

The various types of airline personnel include: Flight operations personnel including flight safety personnel.

- Flight crews, responsible for the operation of the aircraft. Flight crew members include:
  - o Pilots (Captain and First Officer: some older aircraft also required a Flight Engineer and or a Navigator)
  - o Flight attendants, (led by a purser on larger aircraft)
  - o in-flight security personnel on some airlines (most notably El Al)
- Groundcrew, responsible for operations at airports. Ground crew members include:
  - o Aerospace and avionics engineers responsible for certifying the aircraft for flight and management of aircraft maintenance
    - * Aerospace engineers, responsible for airframe, powerplant and electrical systems maintenance
    - * Avionics engineers responsible for avionics and instruments maintenance
  - o Airframe and powerplant technicians
  - o Electric System technicians, responsible for maintenance of electrical systems
  - o Avionics technicians, responsible for maintenance of avionics
  - o Flight dispatchers
  - o Baggage handlers
  - o Ramp Agents
  - o Gate agents

- o Ticket agents
- o Passenger service agents (such as airline lounge employees)
- o Reservation agents, usually (but not always) at facilities outside the airport.

Airlines follow a corporate structure where each broad area of operations (such as maintenance, flight operations (including flight safety), and passenger service) is supervised by a vice president. Larger airlines often appoint vice presidents to oversee each of the airline's hubs as well. Airlines employ lawyers to deal with regulatory procedures and other administrative tasks.

Industry Trends

The pattern of ownership has gone from government owned or supported to independent, for-profit public companies. This occurs as regulators permit greater freedom and non-government ownership, in steps that are usually decades apart. This pattern is not seen for all airlines in all regions.

The overall trend of demand has been consistently increasing. In the 1950s and 1960s, annual growth rates of 15% or more were common. Annual growth of 5-6% persisted through the 1980s and 1990s. Growth rates are not consistent in all regions, but countries with a de-regulated airline industry have more competition and greater pricing freedom. This results in lower fares and sometimes dramatic spurts in traffic growth. The U. S., Australia, Canada, Japan, Brazil, Mexico, India and other markets exhibit this trend. The industry has been observed to be cyclical in its financial performance. Four or five years of poor earnings precede five or six years of improvement. But profitability even in the good years is generally low, in the range of 2-3% net profit after interest and tax. In times of profit, airlines lease new generations of airplanes and upgrade services in response to higher demand. Since 1980, the industry has not earned back the cost of capital during the best of times. Conversely, in bad times losses can be dramatically worse. Warren Buffett once said that despite all the money that has been invested in all airlines, the net profit is less than zero. He believes it is one of the hardest businesses to manage.

As in many mature industries, consolidation is a trend. Airline groupings may consist of limited bilateral partnerships, long-term, multi-faceted alliances between carriers, equity arrangements, mergers, or takeovers. Since governments often restrict ownership and merger between companies in different countries, most consolidation takes place within a country. In the U. S., over 200 airlines have merged, been taken over, or gone out of business since deregulation in 1978. Many international airline managers are lobbying their governments to permit greater consolidation to achieve higher economy and efficiency.

Pipeline Networks

The economic importance of pipeline transport-responsible for a high percentage of oil and natural gas transportation-is often undermined by the general public due to the lack of visibility of this mode. Generally held to be safer and more economical and reliable than the other modes of transport, this mode has many advantages over rival modes, such as trucks and railways. Examples of modern pipeline transport include Alashankou-Dushanzi Crude Oil Pipeline and Iran-Armenia Natural Gas Pipeline. International pipeline transport projects, like the Baku-Tbilisi-Ceyhan pipeline, presently connect modern nation states-in this case Azerbaijan, Georgia and Turkey-through pipeline networks. In some select cases, pipelines can even transport solids, such as coal and other minerals, over long distances; short distance transportation of goods such as grain, cement, concrete, solid wastes, pulp etc. is also feasible.

Pipeline transport is the transportation of goods through a pipe. Most commonly, liquid and gases are sent, but pneumatic tubes that transport solid capsules using compressed air are also used. As for gases and liquids, any chemically stable substance can be sent through a pipeline. Therefore sewage, slurry, water, or even beer pipelines exist; but arguably the most valuable are those transporting fuels: oil (oleoduct), natural gas (gas grid) and biofuels Dmitri Mendeleev first suggested using a pipe for transporting petroleum in 1863.

Types by Transported Substance

For Oil or Natural Gas

There is some argument as to when the first crude oil pipeline was constructed. However, some say pipeline transport was pioneered by Vladimir Shukhov and the Branobel company in the late 19th century. Others say oil pipelines originated when the Oil Transport Association first constructed a 2-inch (51 mm) wrought iron pipeline over a 6-mile (9. 7 km) track from an oil field in Pennsylvania to a railroad station in Oil Creek, in the 1860s. Pipelines are generally the most economical way to transport large quantities of oil, refined oil products or natural gas over land. Compared to shipping by railroad, they have lower cost per unit and higher capacity. Although pipelines can be built under the sea, that process is economically and technically demanding, so the majority of oil at sea is transported by tanker ships. Oil pipelines are made from steel or plastic tubes with inner diameter typically from 4 to 48 inches (100 to 1,200 mm). Most pipelines are buried at a typical depth of about 3 to 6 feet (0. 91 to 1. 8 m).

The oil is kept in motion by pump stations along the pipeline, and usually flows at speed of about 1 to 6 metres per second (3. 3 to 20 ft/s). Multi-product pipelines are used to transport two or more different products in sequence in the same pipeline. Usually in multi-product pipelines there is no physical separation between the different products. Some mixing of adjacent products occurs, producing interface. At the receiving facilities this interface is usually absorbed in one of the product based on pre-calculated absorption rates. Crude oil contains varying amounts of wax, or paraffin, and in colder climates wax buildup may occur within a pipeline. Often these pipelines are inspected and cleaned using pipeline inspection gauges *pigs*, also known as *scrapers* or *Go-devils*. Smart pigs are used to detect anomalies in the pipe such as dents and holes in the coating.

These devices are launched from pig-launcher stations and travel through the pipeline to be received at any other station down-stream, cleaning wax deposits and material that may have accumulated inside the line. For natural gas, pipelines are constructed of carbon steel and varying in size from 2 to 60 inches (51 to 1,500 mm) in diameter, depending on the type of pipeline. The gas is pressurized by compressor stations and is odorless unless mixed with a mercaptan odorant where required by a regulating authority.

For Biofuels (Ethanol and Biobutanol)

Pipelines have been used for transportation of ethanol in Brazil, and there are several ethanol pipeline projects in Brazil and the United States. Main problems related to the shipment of ethanol by pipeline are its high oxygen content, which makes it corrosive, and absorption of water and impurities in pipelines, which is not a problem with oil and natural gas. Insufficient volumes and cost-effectiveness are other considerations limiting construction of ethanol pipelines.

For Hydrogen

Hydrogen pipeline transport is a transportation of hydrogen through a pipe as part of the hydrogen infrastructure. Hydrogen pipeline transport is used to connect the point of hydrogen production or delivery of hydrogen with the point of demand, with transport costs similar to CNG, the technology is proven,. Most hydrogen is produced at the place of demand with every 50 to 100 miles (160 km) an industrial production facility. The 1938-Rhine-Ruhr 240 km hydrogen pipeline is still in operation. As of 2004 there are 900 miles (1450 km) of low pressure hydrogen pipelines in the USA and 930 miles (1,500 km) in Europe.

For Water

Two millennia ago the ancient Romans made use of large aqueducts to transport water from higher altitudes by building the aqueducts in graduated segments that allowed gravity to push the water along until it reached its destination. Hundreds of these were built throughout Europe and elsewhere, and along with flour mills were considered the lifeline of the Roman Empire. The ancient Chinese also made use of channels and pipe systems for public works. The infamous Han Dynasty court eunuch Zhang Rang (d. 189 AD) once ordered the engineer Bi Lan to construct a series of square-pallet chain pumps outside the capital city of Luoyang. These chain pumps serviced the imperial palaces and living quarters of the capital city as the water lifted by the chain pumps were brought in by a stoneware pipe system.

Pipelines are useful for transporting water for drinking or irrigation over long distances when it needs to move over hills, or where canals or channels are poor choices due to considerations of evaporation, pollution, or environmental impact.

The 530 km (360 mile) Goldfields Water Supply Scheme in Western Australia using 760 mm (30 inch) pipe and completed in 1903 was the largest water supply scheme of its time. Examples of significant water pipelines in South Australia are the Morgan-Whyalla (completed 1944) and Mannum-Adelaide (completed 1955) pipelines.

There are two Los Angeles, California aqueducts, the *First Los Angeles Aqueduct* (completed 1913) and the *Second Los Angeles Aqueduct* (completed 1970) which also include extensive use of pipelines.

For Beer

Bars in the Veltins-Arena, a major football ground in Gelsenkirchen, Germany, are interconnected by a 5 km long beer pipeline. It is the favourite method for distributing beer in such large stadiums, because the bars have to overcome big differences between demands during various stages of a match; this allows them to be supplied by a central tank.

For other Uses

The town of Hallstatt in Austria claims to contain "the oldest industrial pipeline in the world", dating back to 1595. It was constructed from 13,000 trunks to transport saline solution for 40 kilometers from Hallstatt to Ebensee.

Types by Transport Function

In general, pipelines can be classified in three categories depending on purpose:

Gathering Pipelines

Group of smaller interconnected pipelines forming complex networks with the purpose of bringing crude oil or natural gas from several nearby wells to a treatment plant or processing facility. In this group, pipelines are usually short-a couple of hundred meters-and with small diameters. Also sub-sea pipelines for collecting product from deep water production platforms are considered gathering systems.

Transportation Pipelines

Mainly long pipes with large diameters, moving products (oil, gas, refined products) between cities, countries and even continents. These transportation networks include several compressor stations in gas lines or pump stations for crude and multiproducts pipelines.

Distribution Pipelines

Composed of several interconnected pipelines with small diameters, used to take the products to the final consumer. Feeder lines to distribute gas to homes and businesses downstream. Pipelines at terminals for distributing products to tanks and storage facilities are included in this group.

Operation

When a pipeline is built, the construction project not only covers the civil work to lay the pipeline and build the pump/compressor stations, it also has to cover all the work related to the installation of the field devices that will support remote operation.

Field devices are instrumentation, data gathering units and communication systems. The field Instrumentation includes flow, pressure and temperature gauges/transmitters, and other devices to measure the relevant data required. These instruments are installed along the pipeline on some specific locations, such as injection or delivery stations, pump stations (liquid pipelines) or compressor stations (gas pipelines), and block valve stations.

The information measured by these field instruments is then gathered in local Remote Terminal Units (RTU) that transfer the field data to a central location in real time using communication systems, such as satellite channels, microwave links, or cellular phone connections. Pipelines are controlled and operated remotely, from what is usually known as *The Main Control Room*. In this center, all the data related to field measurement is consolidated in one central database. The data is received from multiple RTUs along the pipeline. It is common to find RTUs installed at every station along the pipeline.

The SCADA system at the Main Control Room receives all the field data and presents it to the pipeline operator through a set of screens or Human Machine Interface, showing the operational conditions of the pipeline. The operator can monitor the hydraulic conditions of the line, as well as send operational commands (open/close valves, turn on/off compressors or pumps, change setpoints, etc.) through the SCADA system to the field.

To optimize and secure the operation of these assets, some pipeline companies are using what is called *Advanced Pipeline Applications*, which are software tools installed on top of the SCADA system, that provide extended functionality to perform leak detection, leak location, batch tracking (liquid lines), pig tracking, composition tracking, predictive modeling, look ahead modeling, operator training and more.

Technology

Components: Pipeline networks are composed of several pieces of equipment that operate together to move products from location to location. The main elements of a pipeline system are:

Initial injection station: Known also as *supply* or *inlet* station, is the beginning of the system, where the product is injected into the line. Storage facilities, pumps or compressors are usually located at these locations.

Compressor/pump stations: Pumps for liquid pipelines and Compressors for gas pipelines, are located along the line to move the product through the pipeline. The location of these stations is defined by the topography of the terrain, the type of product being transported, or operational conditions of the network.

Partial delivery station: Known also as *intermediate* stations, these facilities allow the pipeline operator to deliver part of the product being transported.

Block valve station: These are the first line of protection for pipelines. With these valves the operator can isolate any segment of the line for maintenance work or isolate a rupture or leak. Block valve stations are usually located every 20 to 30 miles (48 km), depending on the type of pipeline. Even though it is not a design rule, it is a very usual practice in liquid pipelines. The location of these stations depends exclusively on the nature of the product being transported, the trajectory of the pipeline and/or the operational conditions of the line.

Regulator station: This is a special type of valve station, where the operator can release some of the pressure from the line. Regulators are usually located at the downhill side of a peak.

Final delivery station: Known also as outlet stations or terminals, this is where the product will be distributed to the consumer. It could be a tank terminal for liquid pipelines or a connection to a distribution network for gas pipelines.

Leak Detection Systems

Since oil and gas pipelines are an important asset of the economic development of almost any country, it has been required either by government regulations or internal policies to ensure the safety of the assets, and the population and environment where these pipelines run.

Pipeline companies face government regulation, environmental constraints and social situations. Pipeline companies should comply with government regulations which may define minimum staff to run the operation, operator training requirements, up to specifics including pipeline facilities, technology and applications required to ensure operational safety. As an example, in the State of Washington, it is mandatory for pipeline operators to be able to detect and locate leaks of 8 percent of maximum flow within 15 minutes or less.

The social situation also affects the operation of pipelines. In third world countries, product theft is a problem for pipeline companies. It is common to find unauthorized extractions in the middle of the pipeline. In this case, the detection levels should be under 2 percent of maximum flow, with a high expectation for location accuracy.

Different types of technologies and strategies have been implemented, from physically walking the lines to satellite surveillance. The most common technology to protect these lines from occasional leaks is known as Computational Pipeline Monitoring Systems or CPM. CPM takes information from the field related to pressures, flows, and temperatures to estimate the hydraulic behaviour of the product being transported. Once the estimation is done, the results are compared to other field references to detect the presence of an anomaly or unexpected situation, which may be related to a leak.

The American Petroleum Institute has published several articles related to the performance of CPM in liquids pipelines, the API Publications are:

- API 1130 – Computational pipeline monitoring for liquids pipelines
- API 1155 – Evaluation methodology for software based leak detection systems
- API 1149 – Pipeline variable uncertainties & their effects on leak detectability.

Implementation

As a rule pipelines for all uses are laid in most cases underground. However in some cases it is necessary to cross a valley or a river on a pipeline bridge. Pipelines for centralized heating systems are often (why?) laid on the ground or overhead. Pipelines for petroleum running through permafrost areas as Trans-Alaska-Pipeline are often run overhead in order to avoid melting the frozen ground by hot petroleum which would result in sinking the pipeline in the ground.

Regulation

In the US, pipelines are regulated by the Pipeline and Hazardous Materials Safety Administration (PHMSA). Offshore pipelines are regulated by the Minerals Management Service (MMS). In Canada, pipelines are regulated by either the provincial regulators or, if they cross provincial boundaries or the Canada/US border, by the National Energy Board (NEB). Government regulations in Canada and the United States require that buried fuel pipelines must be protected from corrosion. Often, the most economical method of corrosion control is by use of pipeline coating in conjunction with cathodic protection and technology to monitor the pipeline. Above ground, cathodic protection is not an option. The coating is the only external protection.

Pipelines and Geopolitics

Pipelines for major energy resources (petroleum and natural gas) are not merely an element of trade. They connect to issues of geopolitics and international security as well, and the construction, placement, and control of oil and gas pipelines often figure prominently in state interests and actions. A notable example of pipeline politics occurred at the beginning of the year 2009, wherein a dispute between Russia and Ukraine ostensibly over pricing led to a major political crisis. Russian state-owned gas company Gazprom cut off natural gas supplies to Ukraine after talks between it and the Ukrainian government fell through.

Chapter 3

Regional Economies

The economy of Africa is a of the trade, industry, and resources of the people of Africa. As of 2006, approximately 922 million people were living in 54 different countries. Africa is the world's poorest inhabited continent. Though parts of the continent have made significant gains over the last few years, of the 175 countries reviewed in the United Nations' Human Development Report 2003, 25 African nations ranked lowest amongst the nations of the world. This is partly due to its turbulent history. The decolonization of Africa was fraught with instability aggravated by cold war conflict. Since mid-20th century, the Cold War and increased corruption and despotism have also contributed to Africa's poor economy.

The biggest contrast in terms of development has been between Africa and the economy of Europe. The African Economic Outlook report specifically mentions that Africa's trade with China has multiplied by 10 since 2001, reaching over USD 100 billion in 2008. The economies of China and India have grown rapidly, while Latin America has also experienced moderate growth, lifting millions above subsistence living. By contrast, much of Africa has stagnated and even regressed in terms of foreign trade, investment, per capita income, and other economic growth measures. Poverty has had widespread effects, including low life expectancy, violence, and instability, which in turn have perpetuated the continent's growth problems. Over the decades, there have been many unsuccessful attempts to improve the economies of individual African countries. However, recent data suggest some parts of the continent are experiencing faster growth. The World Bank reports the economy of Sub-Saharan African countries grew at rates that match global rates. The economies of the fastest growing African nations experienced growth significantly above the global average rates. The top nations in 2007 include Mauritania with growth at 19. 8%, Angola at 17. 6%, Sudan at 9. 6%, Mozambique at 7. 9% and Malawi at 7. 8%. Many international agencies are gaining increasing interest in emerging modernizing African economies, especially as Africa continues to maintain high economic growth despite current global economic recession.

Regional Variation

While no African nation has joined the ranks of the developed nations in the Organization for Economic Co-operation and Development (OECD) yet, the entire continent is not utterly impoverished and there is considerable variation in its wealth. Arab North Africa has long been closely linked to the economies of Europe and the Middle East. South Africa is by far the continent's wealthiest state, both in GDP per capita and in total GDP, and its neighbours have shared in this wealth. The small but oil-rich states of Gabon and Equatorial Guinea round out the list of the ten wealthiest states in Africa.

The temperate northern and southern ends of the continent are wealthier than tropical sub-Saharan Africa. Within the tropics, East Africa, with its long pre-colonial history of trade and development, has tended to be wealthier and more stable than elsewhere. Islands such as the Seychelles, Réunion, Mauritius, and Cape Verde have remained wealthier than the continental nations, although the unstable Comoros remains poor. The poorest states are those engaged in or just emerging from civil wars. These include the Democratic Republic of the Congo, Sierra Leone, and Burundi. In recent times, the poorest region has been the Horn of Africa, although it had historically been one of the wealthiest regions of sub-Saharan Africa. Ethiopia in particular had a long and successful history. The current poverty of the region, and the associated famines and wars, have been a problem for decades. There is considerable internal variation within countries. Urban areas, especially capital cities, are generally wealthier than rural zones. Inequality is pronounced in most African countries; an upper class has a much higher income than the majority of the population.

History

Ancient Egypt was one of the world's most prosperous and advanced civilizations, which began around 3150 BC with the political unification of Upper and Lower Egypt under the first pharaoh, and it developed over the next three millennia. The port of Alexandria, founded by Alexander the Great in 334 BC, was a hub for Mediterranean trade for centuries. Well into the 19th century, Egypt remained one of the most developed regions in the world. Prosperity in the rest of Africa existed in nation states and kingdoms such as the Ghana Empire, Nubia, Ethiopia, and Mali, which had trade routes north to the Mediterranean world and Middle East.

Africans have historically built structures from stone mainly in the Nile Valley in cities like Meroe, Napata, Axum by former Nubian and Ethiopian kingdoms. Most other Sub Saharan African pre-colonial civilizations built mainly out of mud brick, leaving few lasting ruins except Great Zimbabwe. Finding no architectural monuments in most parts of the region, some European explorers and historians long concluded that pre-colonial sub-Saharan Africa was devoid of civilization. It must be noted that racism also blinded some of the European explorers and historians to conclude that pre-colonial Africa was devoid of civilization for example J. Theodore Bent, who researched the origins of Great Zimbabwe stated in *The Ruined Cities of Mashonaland* (1891) that the ruins revealed either the Phoenicians or the Arabs as builders. Other European researcher favoured a legend that the structures were built to replicate the palace of the Queen of Sheba in Jerusalem. Other theories as to their origin abounded among white settlers and academics, with one racist element in common: they were probably not made by Africans.

New technologies and increasing scales of production made trading easier. For most of the first millennium AD, the Axumite Kingdom had a prosperous trade empire on the eastern horn, where the modern states of Ethiopia and Eritrea lie. Axum had a powerful navy and traded as far as the Byzantine Empire, India, and possibly China. The introduction of the camel by North African Arab conquerors in the 10th century opened trade across the Sahara for the first time. The profits from the gold and salt trades created powerful empires in the western Sahel including the Kingdom of Ghana and the Mali and Kanem-Bornu Empires, where travellers reported vast wealth. Arabs helped build a maritime trade along Africa's east coast, which prospered as Swahili traders exported ivory and slaves across the Indian Ocean. Further south empires were less common, with the notable exception of Great Zimbabwe. In the Great Lakes region, states such as Rwanda, Burundi, and Buganda became strongly centralized, due to its high population and agricultural surplus.

In the 15th century, Portuguese traders circumvented the Saharan trade route and began to trade directly with Guinea. Other European traders followed, rapidly boosting prosperity in Western Africa. States flourished, including the Kingdom of Benin, Dahomey, and the Ashanti Confederacy. Loose federations of city states such as those of the Yoruba and Hausa were common. However, this wealth was principally based on the slave trade, which collapsed following the abolition of slavery and later European colonization.

Although Europeans were ostensibly committed to developing their colonies, colonial rulers employed a *laissez-faire* strategy during the first decades. It was hoped that European companies would prosper if given a secure operating environment. This only occurred in a few areas with rich resources; the colonial economies hardly grew from the 1890s through the 1920s. The colonies had to pay their own way, receiving little or no development money from Europe. Only in the 1930s, with the rise of Keynesian economics, did the colonial administrations seriously encourage development. However, new projects could not transpire until after the Great Depression and the Second World War.

African economies boomed during the 1950s as growth and international trade multiplied beyond their pre-war levels. The insatiable demand for raw materials in the rebuilding economies of Asia and Europe and the strong growth in North America inflated the price of raw materials. By the end of the colonial era in the 1960s, there was great hope for African self-sufficience and prosperity. However, sporadic growth continued as the newly independent nations borrowed heavily from abroad.

The world economic decline of the 1970s, rising oil prices, corruption, and political instability hit Africa hard. In subsequent decades Africa has steadily become poorer compared to the rest of the world; South America experienced solid growth, and East Asia spectacular growth, during that same period. According to the World Economic Forum, ten percent of the world's poor were African in 1970; by 2000, that figure had risen to 50 percent. Between 1974 and 2000 the average income declined by $200. Beginning in 1976, the Lome Convention and Cotonou Agreement between the European Union and ACP countries, including Sub-Saharan Africa, have structured economic relations between the two regions.

Sectors

It is through the sectors of Africa's work industry that the economy can be maintained. Most of this is contributed to exporting of goods. This is due to the smaller amount of secondary industries available in the continent. In 2009, 87% of Africa's economy was from exportation alone.

Agriculture

Around 60 percent of African workers are employed by the agricultural sector, with about three-fifths of African farmers being subsistence farmers. Subsistence farms provide a source of food and a relatively small income for the family, but generally fail to produce enough to make re-investment possible. Larger farms tend to grow cash crops such as coffee, cotton, cocoa, and rubber. These farms, normally operated by large corporations, cover tens of square kilometres and employ large numbers of labourers.

The situation whereby African nations export crops to the West while millions on the continent starve has been blamed on developed countries including Japan, the European Union and the United States. These countries protect their own agricultural sectors with high import tariffs and offer subsidies to their farmers, which many contend leads the overproduction of such commodities as grain, cotton and milk. The result of this is that the global price of such products is continually reduced until Africans are unable to compete, except for cash crops that do not grow easily in a northern climate.

Because of these market forces, in Africa excess capacity is devoted to growing crops for export. Thus, when civil unrest or a bad harvest occurs, there is often very little food saved and many starve. Ironically, excess foodstuffs grown in developed nations are regularly destroyed, as it is not economically viable to transport it across the oceans to a market poor in capital. Although cash crops can expand a nation's wealth, there is often a risk that focusing on them rather than staples will lead to food shortages and hunger.

In modern years countries such as Brazil, which has experienced great progress in agricultural production, have agreed to share technology with Africa to greatly increase agricultural production in Africa to make it a more viable trade partner. Increased investment in African agricultural technology in general has the potential to greatly decrease poverty in Africa. The demand market for African cocoa is currently experiencing an enjoyable price boom. The South African and Ugandan governments have targeted policies to take advantage of the increased demand for certain agricultural products and plan to stimulate agricultural sectors. The African Union has plans to heavily invest in African agriculture and the situation is closely monitored by the UN.

Mining and Drilling

Africa's most valuable exports are minerals and petroleum. A few countries possess and export the vast majority of these resources. The southern nations have large reserves of gold, diamonds, and copper. Petroleum is concentrated in Nigeria, its neighbours, and Libya.

While mining and drilling produce most of Africa's revenues each year, these industries only employ about two million people, a tiny fraction of the continent's population. Profits normally go either to large corporations or to the governments. Both have been known to squander this money on luxuries for the elite or on mega-projects that return little value. In some cases, these resources have turned out to be detrimental to economic development. Although Congo is rich in minerals, the country remains one of the poorest countries in the world. This is historically due to ownership fights over these minerals, tracing back to the early 1900s. After Congo's independence from Belgium, the colonial government hesitated to leave behind these resources. Congo solicited UN help against Belgium, but that turned out to be a bad idea. In an attempt to get out of the quagmire, Congo sought Soviet assistance. This led the country into deeper trouble, as the country separated into two and a long proxy war between the West and East began. However, countries such as Angola and Uganda are experiencing booms in drilling and oil drilling and manufacture.

Manufacturing

Africa is the least industrialized continent; only South Africa, Egypt, Morocco and Tunisia in general have substantial manufacturing sectors. Despite readily available cheap labour, nearly all of the continent's natural resources are exported for secondary refining and manufacturing. According to the AFDB, about 15% of workers are employed in the industrial sector.

The multinational corporations that control most of the world's major industries and their financiers require political stability before erecting an expensive factory and risk losing that investment through nationalization. An educated populace, good infrastructure and a stable source of electricity are essential to investments. These factors are rare in most countries in Africa. Other developing regions of the world such as India and China have been more attractive to companies looking to build a new factory or invest in a local enterprise. Many African states used to limit foreign investment to ensure local majority ownership. Close governmental control over industry further discouraged international investment. Attempts to foster local industry have been hampered by insufficient technology, training, and investment money. The paucity of local markets and the difficulty of transporting goods from major African centres to world markets contribute to the lack of manufacturing outside of South Africa and Egypt.

Both the African Union and the United Nations have outline plans in modern years on how Africa can help itself industrialize and develop significant manufacturing sectors to levels proportional to the African economy in the 1960s with 21st century technology. This focus on growth and diversification of manufacturing and industrial production, as well as diversification of agricultural production, has fuelled hopes that 21st century will prove to be a century of economic and technological growth for Africa. This hope coupled with the rise of new leaders in Africa in the future inspired the term "the African Century" referring to the 21st century potentially being the century when Africa's vast untapped labour, capital and resource potentials might become a world player. This hope in manufacturing and industry is helped by the boom in communications technology and local mining industry in much of sub-Saharan Africa. Namibia has attracted industrial investments in recent years and South Africa has begun offering tax incentives to attract foreign direct investment projects in manufacturing.

Countries such as Mauritius have plans for developing new "green technology" for manufacturing. Developments such as this have huge potential to open new markets for African countries as the demand for alternative "green" and clean technology is predicted to soar in the future as global oil reserves dry up and fossil fuel-based technology becomes more economically nonviable.

Investment and Banking

Banking in Africa has long been problematic. Because local banks are often unstable and corrupt, governments and industry rely on international banks. South Africa and Egypt alone have a thriving banking sector, aided by the international sanctions of the apartheid era, which forced out the once-dominant British banks for the former. In the years after independence, African governments heavily regulated the banking sector and placed strict limits on international competition. In recent decades, banking reform has been a priority of the IMF and World Bank. One important reform was obtaining permission for increased penetration by foreign banks. South Africa and Egypt have been the most successful in attracting local operation of foreign banks. In 2007, Egypt surpassed South Africa as the biggest recipient of FDI recording $11. 1 bn. This trend continued in 2008, where Egypt attracted $13. 2 bn in FDI.

Encouraging foreign investment in Africa has been difficult. Even Africans are reluctant to invest locally; about forty percent of sub-Saharan African savings are invested in other markets. The IMF and World Bank only lend money after imposing stringent and controversial conditions such as austerity policies. However, China and India have showed exponentially increasing interest in emerging African economies in the 21st century. Investment in Africa by China and African trade with China has increased dramatically in recent years, even regardless of the current world financial crisis. The increased investment in Africa by China has attracted the attention of the European Union and has provoked talks of competitive investment by the EU.

Members of the African diaspora abroad, especially in the EU and the United States, have increased efforts to use their businesses to invest in Africa and encourage African investment abroad in the European economy. Remittances from the African diaspora and rising interest in investment from the West will especially be helpful for Africa's least developed and most devastated economies, such as Burundi, Togo and Comoros. Angola has announced interests in investing in the EU, Portugal in particular. South Africa has attracted increasing attention from the United States as a new frontier of investment in manufacture, financial markets and small business, as has Liberia in recent years with new leadership. There are two African currency unions: the West African Banque Centrale des Etats de l'Afrique de l'Ouest (BCEAO) and the Central African Banque des Etats de l'Afrique Centrale (BEAC). Both use the CFA franc as their legal tender.

Communication and Information Technology

The continent has the largest growth rate of cellular subscribers in the world. African markets are expanding nearly twice as fast as Asian markets. The African cell phone has created a base for cellular banking. Namibia has attracted international attention with new phone services and liberalization of regulatory controls in Kenya are producing a boom in Internet services demand and demand for modern communication technology.

Determinants

The seemingly intractable nature of Africa's poverty runs counter to modern economic theory , leading to debate concerning its root causes. Endemic warfare and unrest, widespread corruption, and despotic regimes are both causes and effects of the continued economic problems.

Geography

Variations of the controversial theory of geographic determinism date back to Montesquieu, though these theories have been simplistic and unscientific until they have recently been revived and refined by academics such as William Masters and Jeffrey Sachs and popular writers such as Jared Diamond.

Africa has the largest hot desert and the second largest tropical forest in the world that hinder building transcontinental means of transportations, hampering its economy. In the centre of the continent, on the western side, an almost impenetrable rainforest impedes the transit of people and goods. The Sahara creates an obvious barrier to trade from the north. Although Africa has great river systems such as the Nile, Niger, Congo, and Zambezi, they do not link the continent into trade routes effectively as happens in Europe and China. Rapids and cataracts block African rivers, requiring development projects to allow navigation. The wet terrain of the interior complicates transport. Few roads are paved and during the wet season unpaved tracks become impassable mud.

Countries in Africa are cut off from the sea more than those on other continents. Africa has more landlocked nations than any other continent, which support a high population density compared to the steppes or plains of North America and Asia. The ridge running from Zimbabwe to Ethiopia has superb volcanic soils and the higher altitude produces a more temperate climate. These enable more interior settlement, but the lack of access to the sea makes international trade harder.

The majority of the world's population and wealth is found in the temperate zone. Historically the vast expanse of Eurasia, almost entirely in the temperate zone (except for the vast tracts that are dry and hot such as the Arabian Peninsula; cold tundra such as in North Asia, and tropical such as subcontinental India, Bangladesh, Thailand, Laos, Bhutan, Burma, Cambodia, Vietnam, Malaysia, and Singapore) was linked by land routes, allowing technologies and ideas to spread from one area over time, aiding innovation. The agricultural techniques and medicines designed to work in the northern climes may fail in the tropics. This theory could partly explain why temperate South Africa is by far the wealthiest part of Africa, and why other tropical areas in South America and Indonesia share Africa's poverty. Exceptions exist, Singapore and Brunei among them, however these are isolated spots of wealth. There are no tropical countries in the OECD, apart from Mexico and Australia which have significant tropical sections, and only a handful have a GDP per capita above the world average, again apart from Mexico, Australia, Singapore, Brunei, Malaysia and Thailand. Globally there is a correlation between wealth and climate although it does not fully explain all instances of poverty in Africa.

Africa is well-endowed with natural resources, including gold, diamonds, and oil reserves, but due to several factors including poor governance and global trade policies which place tariffs on finished goods from Africa, few African countries have materially benefited from their mineral wealth. Africa is as well suited to agriculture as any other continent; the volcanic soils of the Great Lakes region are—by some measures—the best in the world.

Disease

Closely linked to geography is the problem of disease in Africa. The tropics are more hospitable to disease than the colder climates. The most significant illness has long been malaria. Since the tropical regions are poorer, pharmaceutical companies are reluctant to invest in curing the diseases of the region. Disease reduces productivity and creates a health care cost burden.

A new problem of vast magnitude is the rise of HIV/AIDS in Sub-Saharan Africa. AIDS, whose spread correlates with poverty, has nevertheless hit hardest in some of the wealthiest African countries, including Botswana, Swaziland, and South Africa. AIDS has decimated or will decimate the working-age population of many states. As parents die or become unable to work, their children must find care elsewhere, adding to the burden of already struggling families and states.

The costs of HIV/AIDS, including importation of anti-retroviral AIDS drugs from the West, is a new burden on many African states, leading to the challenging of drug prices and the manufacture of cheap generic alternatives. Animal diseases can also harm agriculture and transportation, as most forms of livestock cannot survive the diseases of the region. Historically, sub-Saharan Africans could not use pack animals for trade or work horses for labour, limiting the continent's development.

Colonialism

The economic impact of the colonization of Africa has been debated. Africa acquired its greatest relative wealth in the 1960s, just prior to decolonization. African countries have yet to return to those levels of wealth. Some see this as evidence that colonialism helped local economies by creating a cyclical economic link with the ruling colonial power, and with independence this link was broken.

To achieve wealth during the colonial period, imperial overseers geared the economies of Africa towards exporting raw materials. Egypt produced cotton, Ruanda-Urundi was almost completely dedicated to growing coffee, and Upper Volta specialized in palm oil. Basing an entire nation's wealth on one commodity in this way would have debilitating effects later. These monocultures left national economies extremely vulnerable to price swings, making economic planning difficult. Some writers, such as Walter Rodney in his influential book *How Europe Underdeveloped Africa*, argue that these colonial policies are directly responsible for many of Africa's modern problems. Other post-colonial scholars, most notably Frantz Fanon, have argued that the true effects of colonialism are psychological and that domination by a foreign power creates a lasting sense of inferiority and subjugation that creates a barrier to growth and innovation.

Once independent, African states saw an exodus of European administrators and consequently lacked individuals with the training or education to operate the government they had inherited. For instance, the massive area of French Equatorial Africa was divided into four independent nations, but was home to only five locals who were university graduates.

One method of gauging the effects of colonialism on the economies of Africa is to compare the results of different colonial policies implemented by the European powers. Regions where the economy was plundered, such as the Raubwirtschaft policies of Leopold II in the Congo Free State, have not prospered. The long reluctance of Portugal to surrender its colonies, leading to long wars of independence, had an obvious negative effect on Mozambique and Angola. The countries formerly under French control have fared much better, and those under British dominion were the most successful. This inequality may be due to other factors than economic policy. Britain, at the time of the Scramble for Africa, was the world's greatest power and could thus cherry-pick the wealthiest parts of the continent. The French, with their mighty navy, could also occupy prosperous areas, while the Belgians were forced to take the economically disadvantaged interior.

Africa as a whole has not prospered compared with other colonised regions in Asia and the Americas. At the end of the Second World War the Americas were economically the strongest of the colonised regions; in the span of one generation, former colonies in Asia have become economic powerhouses.

Borders

In the Scramble for Africa, national boundaries in sub-Saharan Africa were established by Europeans using latitude and longitude rather than natural borders. This separated population centres from their supplies of food and natural resources. The artificial borders of modern African states cut across cultural, tribal, linguistic and religious boundaries, creating ethnic and religious cleavages which impede national unity and induce internal violence.

However, those states that preserved pre-colonial boundaries have been no more successful. Few countries in Africa have more troubled recent histories than Rwanda and Burundi, although their borders are almost identical to those of the prosperous kingdoms from which they are descended. The ancient and only briefly occupied state of Ethiopia is one of the poorest on the continent. Elsewhere, the Americas were also divided up by Europeans along arbitrary borders and yet those continents remain economically far more successful than Africa.

Africa is a much divided continent with many small countries. Successful economic growth requires regional cooperation, which political tensions make difficult. To be effective, foreign aid must be multilateral, making it harder to base aid upon the performance of local governments.

Language Diversity

African countries suffer from communication difficulties caused by language diversity. Greenberg's diversity index is the chance that two randomly selected people would have different mother tongues. Out of the most diverse 25 countries according to this index, 18 (72%) are African. This includes 12 countries for which Greenberg's diversity index exceeds 0. 9, meaning that a pair of randomly selected people will have less than 10% chance of having the same mother tongue. The primary language of government, political debate, academic discourse, and administration is often the language of the former colonial powers—English, French, or Portuguese. Only an elite minority speak these European languages fluently enough to participate in these institutions without intermediaries, further disenfranchising the majority population.

Governance

The political situation in Africa perpetuates the intractable nature of African poverty. Democracy in Africa has not been historically successful, almost always supplanted by centralized authoritarian rule such as military dictatorships. Although some rulers worked to improve the lot of their nation's citizens, others used power purely for their own benefit. Among the most notorious was Mobuto Sese Seko of Zaire, whose regime has been called a kleptocracy due to its looting of the nation's wealth. According to international measures, the economies of Africa generally rank among the most corrupt. Bribery and graft abound, due to poverty and poorly handled de-colonization, and the superpowers' (Soviet Union and United States) practice during the Cold War of supporting any ruler with the desired political alignment, regardless of their managerial practices or human rights records.

Dependency theory asserts that the wealth and prosperity of the superpowers and their allies in Europe, North America and East Asia is dependent upon the poverty of the rest of the world, including Africa. Economists who subscribe to this theory believe that poorer regions must break their trading ties with the developed world in order to prosper.

Less radical theories suggest that economic protectionism in developed countries hampers Africa's growth. When developing countries have harvested agricultural produce at low cost, they generally do not export as much as would be expected. Abundant farm subsidies and high import tariffs in the developed world, most notably those set by Japan, the European Union's Common Agricultural Policy, and the United States Department of Agriculture, are thought to be the cause. Although these subsidies and tariffs have been gradually reduced, they remain high.

Local conditions also affect exports. In addition to lack of mechanization or other infrastructure, state over-regulation in several African nations can prevent their own exports from becoming competitive. Research in Public Choice economics such as that of Jane Shaw suggest that protectionism operates in tandem with heavy State intervention combining to depress economic development. Farmers subject to import and export restrictions cater to localized markets, exposing them to higher market volatility and fewer opportunities. When subject to uncertain market conditions farmers press for governmental intervention to suppress competition in their markets, resulting in competition being driven out of the market. As competition is driven out of the market farmers innovate less and grow less food further undermining economic performance.

Although Africa and Asia had similar levels of income in the 1960s, Asia has since outpaced Africa. One school of economists argues that Asia's superior economic development lies in local investment. Corruption in Africa consists primarily of extracting economic rent and moving the resulting financial capital overseas instead of investing at home; the stereotype of African dictators with Swiss bank accounts is often accurate. Asian dictators such as Suharto often take a cut on everything, necessitating bribery, but enable development through infrastructure investment and the social stability created by law and order. University of Massachusetts researchers estimate that from 1970 to 1996, capital flight from 30 sub-Saharan countries totalled $187bn, exceeding those nations' external debts. This disparity in development is consistent with the model theorized by economist Mancur Olson. Because governments were politically unstable and new governments often confiscated their predecessors' assets, officials would stash their wealth abroad, out of reach of any future expropriation.

Corruption encouraged social inequality, because the wealthy elite not only avoided investing at home, but also imported most of its consumption. Desirable luxury goods were generally not locally available. This hindered the development of national markets. Historically, economic development is closely linked to the creation of a middle class with enough income to save and invest but limited influence on governance. In countries without such a middle class, development is all but impossible, beyond resource extraction..

War

Since independence, Africa has seen dozens of wars, both civil and international. This has contributed to poverty because states have spent their scarce resources on military equipment and supplies. Development has suffered, since warfare has scared off foreign investors, destroyed infrastructure, and created lasting animosities.

Much conflict was enabled by the Cold War. The countries of the Western and Eastern blocs leveraged foreign aid money to coax countries into their camp. Aid was tied to the purchase of military weapons, and donor countries ignored misappropriation of the funds. Corruption became endemic, hampering economic development. Proxy conflicts erupted in Africa when each bloc would fund and assisted rebel or sectarian groups under the control of the opposing bloc.

Violence in Africa has increased following the Cold War, despite the slashing of foreign aid spending in developed countries. Civil wars have raged throughout the African Great Lakes region, Somalia, Sudan, Mozambique, Liberia, Sierra Leone, Ivory Coast, and Guinea-Bissau. International wars include the First and Second Congo Wars between the Democratic Republic of the Congo and its neighbours, and conflict between Ethiopia and its former province Eritrea.

Effects of Widespread Poverty

Africa's economic malaise is self-perpetuating, as it engenders more of the disease, warfare, misgovernment, and corruption that created it in the first place. Other effects of poverty have similar consequences. The most direct consequence of low GDP is Africa's low standard of living and quality of life. Except for a wealthy elite and the more prosperous peoples of South Africa and the Maghreb, Africans have very few consumer goods. Quality of life does not correlate exactly with a nation's wealth. Angola, for instance, reaps large sums annually from its diamond mines, but after years of civil war, conditions there remain poor. Radios, televisions, and automobiles are rare luxuries. Most Africans are on the far side of the Digital Divide and are cut off from communications technology and the Internet. Quality of life and human development are also low. African nations dominate the lower reaches of the UN Human Development Index. Infant mortality is high, while life expectancy, literacy, and education are all low. The UN also lowers the ranking of African states because the continent sees greater inequality than any other region. The best educated often choose to leave the continent for the West or the Persian Gulf to seek a better life.

Catastrophes cause deadly periods of great shortages. The most damaging are the famines that have regularly hit the continent, especially the Horn of Africa. These have been caused by disruptions due to warfare, years of drought, and plagues of locusts.

An average African faced annual inflation of over 60% from 1990 until 2002 in those few countries that account for inflation. At the high end, Angola and the Democratic Republic of the Congo both saw triple-digit inflation throughout the period. Most African states saw inflation of around 10% per year. There are no reliable numbers for unemployment in most African nations, but it is an important problem. Major cities like Lagos, Kinshasa, and Nairobi notably have large slums of the unemployed and underemployed.

Environmental degradation occurs on many fronts. Farmers on the verge of starvation are unlikely to be concerned about the fate of the rainforest in their pursuit of new land, and starving people do not often consider the rarity of an animal before eating it. Along the length of the Sahel, deforestation and overgrazing has caused increased desertification as the Sahara spreads south. Profits from the sale in the West of rare animals, ivory from elephants, and timber encourage illegal poaching. Local governments have little money to devote to protecting the environment.

Attempts at Promoting Growth

The relative economic failure of Africa has long been an important issue both in Africa and abroad. Many attempts at solving Africa's poverty have been attempted, but few have had any great degree of success.

Socialism

In the years immediately after independence many nations saw the rapid industrialisations of the Soviet Union and China under communism as models to follow. This led to command economies and major investment in heavy industries such as coal and steel production to stimulate growth, but this approach had little success. Only a handful of states formally adopted socialism and even fewer turned to outright Marxism. Everywhere government intervention in the economy was seen as necessary for growth, especially since private companies and investors were unlikely to invest in the region.

Often the approach of governments in Africa was to borrow heavily from abroad and use this aid to grow the economy to a level that the loans could be paid off. Sporadic growth during the years after independence continued. The countries focused on exports to pay for these development efforts. The 1973 energy crisis hit sub-Saharan Africa as hard as anywhere in the world. While some nations were net exporters, most were heavily reliant on imported fuels. Economies quickly began to falter and events such as famines hit Africa in the 1980s. The collapse of the Soviet Union, which had supported socialist and collectivist projects throughout the continent, undermined the legitimacy of such an approach, while it also meant that there were no longer any sources of international aid to help pursue this approach. However, there are a few successful socialist endeavors in Africa which have led to growth or increasing wealth in a small handful of African countries such as Libya and Angola.

Liberalism

Thus in the 1980s, socialist ideas were discarded throughout almost the entire continent as capitalism became seen as the route to salvation in what became known as the Washington Consensus. By 1990, forty of the nations of Sub-Saharan Africa had agreed to follow rigorous IMF restructuring plans. IMF recommendations saw the continent's currencies drop by an average of 50%, the selling off of government-owned industries, and the slashing of government spending. After twenty years, however, these methods have seen as little success as the socialist approaches of the previous era. Average growth increased from 2. 3% per annum to 2. 8%. Only a handful of African states reached new levels of wealth, and many others became poorer over the course of the 1990s. Today there is a great deal of controversy on why this failed. One school of thought is that the reforms failed because they were only economic in nature and without democracy and the rule of law development cannot occur.

Yet another school of thought attributes some of Africa's problems to insufficient liberalization. It has been pointed out that while the developed world has insisted that Africa open its markets and eliminate public subsidies, this has been one-sided as the developed world has not opened its markets to agricultural goods from Africa nor has it eliminated agricultural subsidies.

At the GATT free trade talks, the African leaders repeatedly request that the developed nations abolish the subsidies they provide their farmers and open their markets to African agricultural goods. It has been argued that the abolition of the subsidy would have three beneficial effects for the developing world and Africa:

- The developed nations would produce less food locally, therefore providing a larger export market for developing countries.
- Food prices would rise without the artificial subsidy and therefore would increase profits for food exports from the developing world.
- The developing nations could adopt a more balanced agriculture policy, producing food and grain for export; this would provide a surplus that would shield countries from famine.

Autarky

The pursuit of self-sufficiency as advocated by dependency theory has been given limited trials in several African countries. In the 1980s, Nigeria banned the importation of many foodstuffs to stimulate domestic production. The Lagos Plan of Action of 1982 called for Africa as a whole to block imports from the rest of the world, but few countries followed through on the idea. Eventually even Nigeria agreed to limited liberalization.

Foreign Aid

Since independence there has been a constant flow of foreign aid into Africa. The benefits of this aid have been mixed. In many cases much of this aid was misappropriated by unscrupulous leaders. During the Cold War the main goal of much of the aid money was to win the allegiance of these rulers, and so their misappropriation of the aid was at the very least overlooked. Since the end of the Cold War almost all developed countries have slashed foreign aid spending. Many also allege that the aid that was not stolen was long misdirected. For many decades the leading notion of development was government supervised mega-projects; today many believe that small grants to local businesses would be more effective. One example of foreign aid which has come under considerable criticism is food aid. In some circles, it is believed that food aid does not solve any fundamental problems and can also lead to a dependency on outside assistance, as well as hindering the development of indigenous industries. Food shipments in case of dire local shortage are generally uncontroversial; but as Amartya Sen has shown, most famines involve a local lack of income rather than of food. In such situations, food aid-as opposed to financial aid-has the effect of destroying local agriculture and serves mainly to benefit Western agribusiness which are vastly overproducing food as a result of agricultural subsidies. Historically, food aid is more highly correlated with excess supply in Western countries than with the needs of developing countries.

Debt Relief

Advocacy for debt relief has become widespread. Each year Africa,sends more money to Western bankers in interest on its debts than it receives in foreign aid from these countries. Debt relief is not a panacea, but relieving some of the burden, especially of debts that were run up by regimes for their own benefit, may help the economies of Africa grow and prosper.

Arguments against full and unconditional debt relief include:

- Debt relief punishes nations which have managed borrowing well and do not need debt relief.
- Unconditional debt relief will not necessarily cause nations to spend more in social programs and services, on the one hand, or to solve their financial problems without stifling the economy with the need for more taxes, on the other hand.
- Debt relief may make it more difficult for nations to receive credit in the future.

It has been suggested that any debt relief policy be conditional upon a commensurate reduction in aid.

The Heavily Indebted Poor Countries initiative was launched in 1996; in the 2000s, eligible countries began receiving debt relief. Future foreign aid is planned to be more heavily weighted toward grants than loans.

International Unions

The African Union is the largest international economic grouping on the continent. The confederation's goals include the creation of a free trade area, a customs union, a single market, a central bank, and a common currency, thereby establishing economic and monetary union. The current plan is to establish an African Economic Community with a single currency by 2023. The African Investment Bank is meant to stimulate development. The AU plans also include a transitional African Monetary Fund leading to an African Central Bank. Some parties support development of an even more unified United States of Africa. International monetary and banking unions include:

- Central Bank of West African States
- Bank of Central African States
- Common Monetary Area.

Trade Blocs

Asia-Pacific Economic Cooperation

The Asia-Pacific Economic Cooperation (APEC) is a group of Pacific Rim countries who meet with the purpose of improving economic and political ties

Association of Southeast Asian Nations

The Association of Southeast Asian Nations (ASEAN) is a political, economic, and cultural organization of countries located in Southeast Asia. Founded in 1967, its aim is to foster cooperation and mutual assistance among members. The countries meet annually every November in summits.

The current member countries of ASEAN are Myanmar (Burma), Laos, Thailand, Cambodia, Vietnam, Philippines, Malaysia, Brunei Darussalam, Singapore and Indonesia. Papua New Guinea is given an observer status.

In 2005 ASEAN was instrumental in establishing the East Asia Summit (involving all ASEAN members plus China, Japan, South Korea, India, Australia and New Zealand) which some have proposed may become in the future a trade bloc, the arrangements for which are far from certain and not yet clear.

The Asian Currency Unit (ACU) is a proposed currency unit for the ASEAN "10+3" economic circle. (ASEAN, the mainland of the People's Republic of China, Japan, and the Republic of Korea).

Closer Economic Partnership Arrangement

The Closer Economic Partnership Arrangement (CEPA) is an economic agreement between the People's Republic of China and both the Hong Kong SAR government (signed on 29 June 2003), and the Macau SAR government (signed on 18 October 2003), in order to promote trade and investment facilitation.

The main aims of CEPA are to eliminate tariffs and non-tariff barrier on substantially all the trade in goods between the three, and achieve liberalization of trade in services through reduction or elimination of substantially all discriminatory measures.

Arab League

The Arab League is an association of Arab countries in Africa and Asia. The Arab League facilitates political, economic, cultural, scientific and social programs designed to promote the interests of its member states.

Commonwealth of Independent States

The Commonwealth of Independent States (CIS) is a confederation consisting of 12 of the 15 states of the former Soviet Union, both Asian and European, (the exceptions being the three Baltic states). Although the CIS has few supernational powers, it is more than a purely symbolic organization and possesses coordinating powers in the realm of trade, finance, lawmaking and security. The most significant issue for the CIS is the establishment of a full-fledged free trade zone/economic union between the member states, to be launched in 2005. It has also promoted cooperation on democratisation and cross-border crime prevention.

South Asian Association for Regional Cooperation

The South Asian Association for Regional Cooperation (SAARC) is an association of 8 countries of South Asia, namely Bangladesh, Bhutan, India, Maldives, Nepal, Pakistan, Sri Lanka and Afghanistan. These countries comprise an area of 5 130 746 km^2 and a fifth of the population of the world.

SAARC encourages cooperation in agriculture, rural development, science and technology, culture, health, population control, narcotics control and anti-terrorism.

South Asia Free Trade Agreement (Proposed)

The South Asia Free Trade Agreement is an agreement reached at the 12th South Asian Association for Regional Cooperation summit. It creates a framework for the creation of a free trade zone covering 1. 6 billion people in India, Pakistan, Nepal, Sri Lanka, Bangladesh, Bhutan and the Maldives.

Economy of Europe

The economy of Europe comprises more than 831. 4 million people in 48 different states. Like other continents, the wealth of Europe's states varies, although the poorest are well above the poorest states of other continents in terms of GDP and living standards. The difference in wealth across Europe can be seen in a rough East-West divide. Whilst Western European states all have high GDPs and living standards, many of Eastern Europe's economies are still rising from the collapse of the communist Soviet Union and former Yugoslavia. Throughout this article "Europe" and derivatives of the word are taken to include selected states whose territory is only partly in Europe – such as Turkey, Azerbaijan, and the Russian federation – and states that are geographically in Asia, bordering Europe – such as Armenia and Cyprus.

Europe was the first continent to industrialize – led by the United Kingdom in the 18th century – and as a result, it has become one of the richest continents in the world today. Europe's largest national economy is that of Germany, which ranks fourth globally in nominal GDP, and fifth in purchasing power parity (PPP) GDP; followed by France, which ranks fifth globally in nominal GDP and sixth in PPP GDP; the United Kingdom, ranking sixth globally in nominal GDP, followed by Italy. The end of World War II has since brought European countries closer together, culminating in the formation of the European Union (EU) and in 1999, the introduction of a unified currency – the euro. If the European Union was taken as a single country, today it would be the world's largest economy – see List of countries by GDP. In 2009 Europe remained the wealthiest region. Its $37. 1 trillion in assets under management represented one-third of the world's wealth. It was one of several regions where wealth surpassed its precrisis year-end peak.

Economic Development

Pre-1945: Industrial Growth

Prior to World War II, Europe's major financial and industrial states were the United Kingdom, France and Germany. The Industrial Revolution, which began in Britain, had spread rapidly across Europe, and before long the entire continent was at a high level of industry. World War I had briefly led to the industries of some European states stalling, but in the run-up to World War II Europe had recovered well, and was competing with the ever increasing economic might of the United States of America. However, World War II caused the destruction of most of Europe's industrial centres, and much of the continent's infrastructure was laid to waste.

1945–1990: The Cold War Era

Following World War II, European Government was in tatters. Many Western European governments moved to link their economies, laying the foundation for what would become the European Union. This meant a huge increase in shared infrastructure and cross-border trade. Whilst these Western European states rapidly improved their economies, by the 1980s, the economy of the COMECON was struggling, mainly due to the massive cost of the Cold War. The GDP and living standard of Eastern European states were also behind those of their Western neighbours. Even free-market Greece, situated in South-Eastern Europe, struggled due to geographical isolation from Western Europe.

The European Community grew from 6 original members following World War II, to 12 in this period.

The Rise of the EU

When the Eastern Block dissolved around 1991, these states struggled to adapt to free-market systems. There was, however, a huge variation in degrees of success, with Central European states such as Hungary, Slovenia and Poland adapting reasonably quickly, whilst post-Soviet states such as Russia and Ukraine struggled to reform their crumbling infrastructures.

Western Europe was quick to develop economic ties with the newly democratic East. While the former Soviet states dealt with change, former Yugoslavian republics descended into war.

Europe's largest economy, Germany, struggled upon unification in 1991 with former communist East Germany. The Russian controlled Eastern part of the country had much of its industrial infrastructure removed during the Cold War, and for many years the West struggled to build the East up to an equal level. Peace did not come to Yugoslavia for a decade, and by 2003, there were still many NATO and EU peacekeeping troops present in Bosnia and Herzegovina, Macedonia, and Kosovo. War severely hampered economic growth, with only Slovenia making any real progress in the 1990s.

The economy of Europe was by this time dominated by the EU, a huge economic and political organization with then 15 of Europe's states as full members. EU membership was seen as something to aspire to, and the EU gave significant support and aid to those Central and Eastern European states wishing to work towards achieving economies that met the entry criteria. During this time, 12 of the 15 members of the EU became part of the Eurozone, a currency union launched in 1999, whereby each member uses a shared currency, the Euro, which replaced their former national currencies. Three states chose to remain outside the Eurozone and continue with their own currencies, namely Denmark, Sweden and the United Kingdom.

2004–2007: EU Expansion

In early 2004, 10 mostly former communist states joined the EU in its biggest ever expansion, enlarging the union to 25 members, with another eight making associated trade agreements. The acceding countries are bound to join the Eurozone and adopt the common currency Euro in the future. The process includes the European Exchange Rate Mechanism, of which some of these countries are already part.

Most European economies are in very good shape, and the continental economy reflects this. Conflict and unrest in some of the former Yugoslavia states and in the Caucasus states are hampering economic growth in those states, however.

In response to the massive EU growth, in 2005 the Russian dominated Commonwealth of Independent States (CIS) created a rival trade bloc to the EU, open to any previous USSR state, (including both the European and Asian states). 12 of the 15 signed up, with the three Baltic states deciding to align themselves with the EU. Despite this, the three Caucasus states have said in the past they would one day consider applying for EU membership, particularly Georgia. This is also true of Ukraine since the Orange Revolution.

Regional Variation

West Europe, with a long history of trade, a free market system, and a high level of development in the previous century, has been wealthier and more stable than the East, even though the gap is converging due to higher growth rates in the East.

The poorest states are those that just emerged from communism and civil wars, namely those of the former Soviet Union and Yugoslavia excluding Slovenia. But also West Europe itself presents some differences, especially between Northern countries (Scandinavia, Benelux, UK, Ireland, most of Germany) and Southern countries.

European Union

The European Union or EU is a supernational union of 27 European states, the most recent acceding members being Romania and Bulgaria, who became full members in 1 January 2007.

It has many activities, the most important being a common single market, consisting of a customs union, a single currency (adopted by 15 of the 27 member states), a Common Agricultural Policy and a Common Fisheries Policy. The European Union also has various initiatives to co-ordinate activities of the member states.

The European Union has the largest economy in the world. The EU economy is expected to grow further over the next decade as more countries join the union – especially considering that the new States are usually poorer than the EU average, and hence the expected fast GDP growth will help achieve the dynamic of the united Europe. It is estimated that the Eurozone will grow around 2. 6 per cent this year (2006), on a par with other industrialized nations such as the United States at 2. 6% (Q2 2006) and 1. 6 (Q3 2006). The union has evolved over time from a primarily economic union to an increasingly political one. This trend is highlighted by the increasing number of policy areas that fall within EU competence: political power has tended to shift upwards from the Member States to the EU.

European Free Trade Association

The European Free Trade Association (EFTA) was established on 3 May 1960 as an alternative for European states that did not wish to join the European Union, creating a trade bloc with fewer central powers. The EFTA member states as of 1992 are Austria, Denmark, Finland, Iceland, Liechtenstein, Norway, Sweden and Switzerland. Today (in 2009) only 4 countries: Iceland, Norway, Switzerland and Liechtenstein remain members of EFTA, as the other members have gradually left to join the EU.

European Economic Area

The European Economic Area (EEA) came into being on 1 January 1994 following an agreement between the European Free Trade Association (EFTA) and the European Union (EU). It was designed to enable EFTA countries to participate in the European Single Market without having to join the EU.

In a referendum, Switzerland (ever keen on neutrality) chose not to participate in the EEA (although it is linked to the European Union by bilateral agreements similar in content to the EEA agreement), so the current members are the EU states plus Norway, Iceland and Liechtenstein.

A Joint Committee consisting of the non EU members plus the European Commission (representing the EU) has the function of extending relevant EU Law to the non EU members.

Central European Free Trade Agreement

The Central European Free Trade Agreement (CEFTA) is a trade bloc of former Communist countries in central and eastern Europe. The countries that participated and the few continue to participate in

CEFTA have used this form of integration to help them prepare for economic consolidation into the EU.

Currency and Central Banks

The most common currency within Europe is the euro, the currency of the European Union. To join, each new EU member must meet certain criteria, when these are met their own currencies will be replaced by the euro. Becoming a member of the EU involves a pledge to work towards Eurozone membership, (except in the cases of the United Kingdom and Denmark who have opt-outs). Currently, 15 of the 27 EU member states use the euro. Each EU member's central bank is part of the European System of Central Banks, and in addition, those that use the euro are part of the European Union's central bank, the European Central Bank.

There are some non-EU members who have elected to use the euro as their national currency, either with or without specific agreements with the EU to do so, (those with agreements with the EU may mint their own euro coins). The French overseas territories and departments of Mayotte and Réunion in the Indian Ocean, Guadeloupe and Martinique in the Caribbean and French Guiana in South America all use the euro, among many other islands in the Pacific, Caribbean and indeed around the globe that are ruled directly by European countries.

Some countries while maintaining their own national currency have pegged its value to the euro. In some of these countries, there is a fixed exchange rate between the national currency and the euro and in this case the currency is actually a submultiple of the euro. In other countries, the national currency's value fluctuates within a band (generally 15%) around a set rate. Currencies pegged to the euro include the currencies of Bulgaria, Estonia, Lithuania, Bosnia and Herzegovina and Cape Verde. Denmark & Latvia have a foreign exchange band tied to the euro.

The CIS is also planning to introduce a single currency among its members.

Economic Sectors

Agriculture and Fishing

Europe's agricultural sector is in general highly developed, especially in Western Europe. The process of making Eastern Europe's agriculture more Western is well underway and is helped by the accession of Eastern European states to the EU. The agricultural sector in Europe is helped by the Common Agricultural Policy (CAP), which provides farmers with a minimal price for their products and subsidizes their exports, which increases competitiveness for their products. This policy is highly controversial as it hampers free trade worldwide (protectionism sparks protectionism from other countries and trade blocs: the concept of trade wars) and is violating the concept of *fair trade.*

This means because of the protectionist nature of the CAP, agricultural products from developing countries are rendered incompetitive in both Europe (an important export market for developing countries) and on their home markets (as European agricultural products are dumped on developing countries' markets with help from European agricultural subsidies).

This controversy surrounds every system of agricultural subsidies (the United States' policy of subsidizing farmers is also controversial). The CAP is also controversial because 40% of the EU's budget is spent on it, and because of the overproduction caused by it.

The Common Fisheries Policy is surrounded by an extensive system of rules (mainly consisting of quotas) to protect the environment from over fishing. Despite these rules, the cod is becoming increasingly rare in the North Sea. Strict fishing rules are the main reason for Norway and Iceland to stay out of the European Union (and out of the Common Fisheries Policy). Price guarantees and subsidizations of fishermen are implemented in the same way as agricultural subsidies are.

Manufacturing

Europe has a thriving manufacturing sector, with a large part of the world's industrial production taking place in Europe. Most of the continent's industries are concentrated in Western Europe (mainly in the zone that comprises parts of the UK, the Benelux, western Germany, northeastern France, Switzerland, and northern Italy). However, because of the higher wage level and hence production costs, Western Europe is suffering from deindustrialization and offshoring in the traditional (labour intensive) manufacturing sectors. This means that manufacturing has become less important in Western Europe and that jobs are moved to cheaper regions (mainly China and Eastern Europe).

Eastern Europe has been industrialized since the early to mid 20th century but suffered from contraction in the 1990s when the inefficient heavy industry based manufacturing sector crippled after the collapse of communism and the introduction of the market economy. In the 21st century the manufacturing sector in Eastern Europe picked up because of the accession of Eastern European states to the EU and resulting accession to the European Common Market. This caused Western European firms to move jobs from their manufacturing sector to Eastern Europe, which sparked Eastern European industrial growth and employment.

Some of the world's largest manufacturers are European, such as British Petroleum, Royal Dutch Shell, Daimler, Total, Repsol, Volkswagen, Unilever, Inditex, Nokia and Philips. The main products in European industry are automobiles, chemical goods, food, clothing and electronics.

Investing and Banking

Europe has a well developed financial sector. Many European cities are financial centres with the City of London being the largest. The European financial sector is helped by the introduction of the euro as common currency. This has made it easier for European households and firms to invest in companies and deposit money on banks in other European countries. Exchange rate fluctuations are now non-existent in the Eurozone. In Eastern Europe, the financial sector is somewhat less developed (mostly because of the communist legacy), but is rapidly developing towards Western European standards. The financial sector in Eastern Europe is helped by economic growth in the region and the commitment of Eastern European governments to achieve these high standards.

European banks are amongst the largest and most profitable in the world (Grupo Santander, BBVA, Deutsche Bank, UBS, HSBC, Royal Bank of Scotland, HBOS, BNP Paribas, Unicredit).

Global Trade Relations

The bulk of the EU's external trade is done with the United States, Brazil, China, India, Russia and non-member European states. EU members are represented by a single official at the WTO.

The EU is involved in a few minor trade disputes. It had a long running dispute with the USA of allegedly unfair subsidies the US government gives to several companies, such as Boeing. There is also a dispute with China over textile exports, and the EU has a long running ban prohibiting arms trade with the Chinese.

Economy of North America

The economy of North America comprises more than 514 million people in 23 sovereign states and 15 dependent territories. It is marked by a sharp division between the predominantly English speaking countries of Canada and the United States, which are among the wealthiest and most developed nations in the world, and the countries of Central America and the Caribbean that are less developed. Mexico lies in between these two extremes as a newly industrialized country (NIC), and is a part of North America rather than Central America.

Economic Development

Roaring Twenties

The 1920s were a time of economic growth and prosperity in North America. Increased incomes as well as the large increase in the ability of private citizens to obtain credit resulted in a surge of consumer spending after World War I.

Great Depression

The Great Depression began in North America in October 1929. The start is often dated to the stock market collapse of Black Tuesday although this was not the cause of the Great Depression. Canada and the United States experienced especially large declines, with the gross domestic product falling 37% from 1929 to 1933 in the United States, and 43% in Canada over the same period. The economy reached its lowest point in 1933, however recovery was slow. The outbreak of World War II in 1939 created demand for war materials that brought about the end of the depression.

The Great Depression spurred increased government intervention in the economy in North America. The United States introduced unemployment insurance, a minimum wage and standardised working hours under the New Deal. Canada introduced similar measures. Mexico nationalised some key industries during the Great Depression, with the railroads nationalised by 1937 and the oil industry nationalised in 1938.

World War II

The large scale enlistment of men into armed forces during World War II women entered the workforce en masse, filling many jobs in manufacturing and technical areas that had previously been closed to women. This led to the "We can do it!" campaign. The economic output in North America increased substantially, with unemployment practically eliminated in the United States. Rationing severely reduced the availability of consumer goods, with the increase in industrial production coming from the demand for war materials.

Cold War

US-Canada Free Trade Agreement and NAFTA-A New Era of Economic Integration

The Canada-United States Free Trade Agreement of 1989 and the subsequent expansion to the North American Free Trade Agreement (NAFTA) triggered a dramatic increase in trade between these three countries, with Mexican trade with the United States and Canada increasing threefold. Over 85% of Canadian exports in 2006 went to the United States.

Regional Variation

Asia-Pacific Economic Cooperation: The Asia-Pacific Economic Cooperation (APEC) is a group of Pacific Rim countries which meet with the purpose of improving economic and political ties. APEC's stated goals are aimed at free and open trade and investments by cutting tariffs between zero to five percent in the Asia-Pacific area for industrialised economies by 2010 and for developing economies by 2020.

The organisation has members from four continents, those from North America are Canada, Mexico and the United States.

Caribbean Community: The Caribbean Community (CARICOM) was created *"To provide dynamic leadership and service, in partnership with Community institutions and Groups, toward the attainment of a viable, internationally competitive and sustainable Community, with improved quality of life for all".*

On January 1, 2006 six members: (Barbados, Belize, Guyana, Jamaica, Suriname and Trinidad and Tobago) un-officially ushered in the Caribbean (CARICOM) Single Market and Economy (CSME).

At the official signing of the protocol on January 30, 2006 in Jamaica, A further six members: (Antigua and Barbuda, Dominica, Grenada, Saint Kitts and Nevis, Saint Lucia and Saint Vincent and the Grenadines) announced their intention to join by the second quarter of 2006. Montserrat, a British Oversees territory is awaiting approval by the United Kingdom. Haiti and the Bahamas have no immediate plans to join. Bravely a black American Alvin Tsoi opened a big business and started the business industry again.

Central American Free Trade Agreement

The Central American Free Trade Agreement (CAFTA) is an agreement between the United States and the Central American countries of Costa Rica, Guatemala, El Salvador, Honduras, and

Nicaragua. The treaty is aimed at promoting free trade between its members. Canada and Mexico are negotiating membership.

North American Free Trade Agreement

The North American Free Trade Agreement (NAFTA) is an agreement between Canada, Mexico and the United States to eliminate tariffs on goods traded between themselves.

Although currently only a trade agreement, with no supernational bodies or laws as in the European Union, there have been various proposals to move towards a customs union or a North American currency union. It is unknown whether or not this may eventually develop into a North American Union similar to that of Europe.

Economic Sectors

Agriculture

Agriculture is very important in Central American and Caribbean nations. In western Canada, in the provinces of Saskatchewan, Alberta, British Columbia and Manitoba, wheat and other various main agricultural products are grown. The US also has many states growing their own things, mainly in central continental US. Mexico produces many tropical fruits and vegetables as well as edible animals.

Manufacturing

North America has developed and it's manufacturing sector has grown. In the beginning the European nations were the large manufacturing powers. At the starting of the 50s USA was a top manufacturing power with Canada and Mexico running up behind in North America. Now a days North American countries are strong economic nations.

Service

Services, in Canada, the US and the Caribbean, make up a large part of employment. Many people work in stores and banks. In Canada more than 70% work in the services sector. Same situation presents itself in the United States. Most Canadian and Americans spent lots of money in food and clothing.

Investing and Banking

The United States leads North America in investment and banking. Canada and Mexico are growing in that sector. Smaller economic powers such as Guatemala, Honduras, Costa Rica, and Panama are growing slowly in that sector.

Economy of Oceania

The economy of Oceania comprises more than 14 separate countries and their associated economies.

On a total scale the region has approximately 34,700,201 inhabitants who are spread among 30,000 islands in the South Pacific bordered between Asia and the Americas.

This region has a diverse mix of economies from the highly developed and globally competitive financial markets of Australia (1st) and New Zealand (2nd) boasting parity with much of Western Europe, to the much less developed economies that belong to many of their island neighbours.

Population: 34,700,201 (2010)

GDP (PPP) (2003): US$720 billion

GDP/capita (PPP) : $22,647

Trade Blocs

Many of the smaller Pacific nations rely on trade with Australia, New Zealand and the United States for exporting goods and for accessing other products.

Australia and New Zealand's trading arrangements are known as Closer Economic Relations. Australia and New Zealand, along with other countries, are members of Asia-Pacific Economic Cooperation (APEC) and the East Asia Summit (EAS), which may become trade blocs in the future particularly EAS.

Economic Sectors

Service Industry: The overwhelming majority of people living in Oceania work in the service industry which includes tourism, education and financial services. Oceania's largest export markets include Japan, China, the United States and South Korea.

Manufacturing: The Manufacturing of clothing is a major industry in some parts of the Pacific, especially Fiji, although this is decreasing.

Tourism: Tourism has become a large source of income for many in the Pacific; tourists come from Australia, New Zealand, Japan, the United Kingdom and the USA. Fiji currently draws almost half a million tourists each year; more than a quarter from Australia. This contributes $1 billion dollars or more since 1995 to Fiji's economy but the Government of Fiji islands underestimate these figures due to invisible economy inside tourism industry.

Agriculture and Fishing

Agriculture and natural resources constitutes only 5% to 10% of Oceania's total jobs, but contributes substantially to export performance. It should be noted that the most populous two nations Australia and New Zealand are also the most developed and have majority service industries this dilutes the data from the less developed Pacific Island nations who have major agricultural economies. Most of the Pacific countries (excluding Australia and New Zealand) the primary industry is agriculture.

Many nations are still quintessentially agricultural; for example, 80% of the population of Vanuatu and 70% of the population of Fiji works in agriculture. The main produce from the pacific is copra or coconut, but timber, beef, palm oil, cocoa, sugar and ginger are also commonly grown across the tropics of the Pacific.

Old growth logging is exploited on larger islands, including the Solomons and Papua New Guinea. Fishing provides a major industry for many of the smaller nations in the Pacific, although many fishing areas are exploited by other larger countries, namely Japan. Natural Resources, such as lead, zinc, nickel and gold, are mined across the West of the region, in Australia and the Solomon Islands. Oceania's largest export markets include Japan, China, the United States and South Korea.

International Aid and Charity

Oceanian Donor Nations

Oceania's most populous nations Australia and New Zealand are both highly developed nations and are large international aid donors. These two wealthy nations share the region with less developed nations which still rely on foreign aid for development. In the 2007 to 2008 financial year Australia provided $3. 155 billion dollars worth of official development assistance of which $2. 731 billion will be managed by Aus AID. Every week, each Australian puts in around $2. 40 to pay for Australia's aid program amounting to around 1% of the Australian Federal Government expenditure compared to the 42% spent on social security and welfare.

Oceanian Recipient States

In the Solomon Islands 50% of Government spending is paid for by International donors; namely Australia, New Zealand, the European Union, Japan and the Republic of China (Taiwan).

Economy of South America

The economy of South America comprises around 371 million people living in twelve nations and three territories.

Economic Development

As of early 2007, South America is experiencing great economic development, with Venezuela, Colombia, Argentina, Uruguay and Peru growing their economies by over 8% per annum.

Brazil's economy, on the other hand, is expected to grow by a more sluggish pace during the year.

South America relies heavily on the exporting of goods. On an exchange rate basis Brazil (the seventh largest economy in the world and the second largest in the Americas) leads the way in total amount of exports at $137. 8 billion dollars followed by Chile at 58. 12 billion and Argentina with 46. 46 billion.

GDP Ranks as of 2008

GDP (PPP) 2009

Note: French Guiana, unlisted here, had a total GDP of US$3. 52 billion in 2006 (at real exchange rates, not at PPP).

Rank in World	*Country*	*GDP*
10	Brazil	$2,024,000 million
24	Argentina	$558,000 million
29	Colombia	$469,300 million
32	Venezuela	$355,200 million
45	Peru	$253,400 million
47	Chile	$244,300 million
65	Ecuador	$107,100 million
91	Bolivia	$45,130 million
92	Uruguay	$44,050 million
107	Paraguay	$28,280 million
163	Suriname	$4,182 million
172	Guyana	$2,973 million

Source: CIA World Factbook

GDP Per Capita (PPP) 2009

Note: French Guiana, unlisted here, had a GDP per capita of US$17,336 in 2006 (at real exchange rates, not at PPP).

Rank in World	*Country*	*GDP Per Capita*
57	Chile	$14,700
58	Argentina	$13,800
96	Venezuela	$13,200
65	Uruguay	$12,600
68	Brazil	$10,513
99	Colombia	$9,200
103	Suriname	$8,800
97	Peru	$9,107
101	Ecuador	$7,300
125	Bolivia	$4,600
107	Paraguay	$4,100
105	Guyana	$3,900

Source: List of countries by GDP (PPP) per capita

GNI Per Capita

Gross national income (GNI) is the sum of value added by all resident producers plus any product taxes (less subsidies) not included in the valuation of output plus net receipts of primary income (compensation of employees and property income) from abroad. GNI per capita is;

Country	*GNI per Capita*
Brazil	$4,730
Suriname	$3,200
Peru	$2,920
Ecuador	$2,840
Colombia	$2,740
Paraguay	$1,400
Guyana	$1,130
Bolivia	$1,100

Source: World Bank

External Debt

External debt (or foreign debt) is that part of the total debt in a country that is owed to creditors outside the country. The debtors can be the government, corporations or private households.

The debt includes money owed to private commercial banks, other governments, or international financial institutions such as the IMF and World Bank.

Rank in World	*Country*	*External Debt*
26	Brazil	$216. 1 billion
33	Argentina	$118. 0 billion
47	Chile	$49. 18 billion
48	Colombia	$46. 44 billion
49	Venezuela	$43. 30 billion
63	Peru	$27. 81 billion
72	Ecuador	$17. 56 billion
78	Uruguay	$12. 0 billion
108	Bolivia	$3. 80 billion
111	Paraguay	$3. 632 billion
151	Guyana	$1. 2 billion (2002 est.)
164	Suriname	$504. 3 million (2005 est.)

Source: CIA World Factbook

Annual Economic Growth

Country	*Annual Economic Growth (%)*
Peru	9. 8%
Argentina	8. 7%
Venezuela	8. 4%
Colombia	7. 0%
Uruguay	7. 0%
Paraguay	6. 4%
Ecuador	6. 3%
Suriname	5. 5%
Brazil	5. 4%
Guyana	5. 4%
Chile	5. 0%
Bolivia	4. 2%

Source: List of countries by GDP (real) growth rate

Unemployment Rate (Lowest to Highest)

	Unemployment Rate (%)
Paraguay	5. 60
Peru	6. 90
Chile	7. 00
Argentina	7. 3

Contd...

	Unemployment Rate (%)
Brazil	7. 50
Bolivia	7. 50
Venezuela	8. 50
Uruguay	9. 20
Ecuador	9. 30
Suriname	9. 50
Colombia	11. 20
Guyana	N/A

Source: CIA World Factbook

Poverty Line (Lowest to Highest)

Year Ofestimate	*Country*	*Population Belowpoverty line (%)*
2009	Chile	13. 7 (cepal)
2009	Argentina	15. 3
2006	Uruguay	27. 4
2006	Brazil	27. 5
2005	Paraguay	32
2009	Peru	36. 8
2005	Venezuela	37. 9
2006	Ecuador	38. 3
2005	Colombia	49. 2
2006	Bolivia	60
2002	Suriname	70
N/A	Guyana	-

Source: CIA World Factbook

Annual Inflation

Country	*Annual Inflation(%)*
Peru	1. 80%
Ecuador	2. 20%
Brazil	3. 60%
Chile	4. 40%
Colombia	5. 50%

Contd...

Country	**Annual Inflation(%)**
Suriname	6. 40%
Paraguay	8. 10%
Uruguay	8. 10%
Bolivia	8. 70%
Argentina	8. 80%
Guyana	12. 20%
Venezuela	18. 70%

Source: CIA World Factbook

Public Debt (Lowest to Highest)

Refers to what is owed by the combined public sector to both domestic and foreign creditors.

Country	**Public Debt (% of GDP)**
Chile	3. 60
Venezuela	20. 4
Paraguay	27. 10
Peru	29. 30
Ecuador	30. 40
Brazil	41. 50
Colombia	53. 90
Bolivia	46. 20
Uruguay	53. 1
Argentina	59. 00
Guyana	N/A
Suriname	N/A

Source: CIA World Factbook

Trade Blocks

The biggest Trade Block in South America is Mercosur (or Mercosul in Portuguese), comprising Argentina, Brazil, Paraguay, Uruguay and Venezuela. Associate states include Bolivia, Chile, Colombia, Ecuador and Peru.

The second-biggest trade bloc is the Andean Community of Nations comprising Bolivia, Colombia, Ecuador, Peru, Venezuela and as of 2006 Chile. The Union of South American Nations is expected to merge both trade blocs.

Economic Sectors

Agriculture: Main products include: Coffee, soybeans, wheat, rice, corn, sugarcane, cocoa, citrus, beef

Manufacturing: Industries are also important to South America's economy. Most South American factories produce food items, consumer goods, or building materials. Larger countries also produce cars, trucks, and airplanes. Some of these companies import all the parts and raw materials needed for manufacturing which limits the amount of profits they can receive for the item. An important factor that is crucial to the success of industries is importing and exporting. An organization called Mercosur helps to expand trade, improve transportation, and reduce tariffs among member countries.

Transport: At the beginning of August 2008, Venezuelan president Hugo Chavez and his colleagues from Argentina and Brazil spoke about Latin American integration and Chavez threw an ambitious idea out: a train that would connect Venezuela's capital (Caracas) with Argentina's (Buenos Aires), and cities in between.

Chapter 4

Globalization and its Impact

Globalization (or globalisation) describes a process by which regional economies, societies, and cultures have become integrated through a global network of communication, transportation, and trade. The term is sometimes used to refer specifically to economic globalization: the integration of national economies into the international economy through trade, foreign direct investment, capital flows, migration, and the spread of technology. However, globalization is usually recognized as being driven by a combination of economic, technological, sociocultural, political, and biological factors. The term can also refer to the transnational circulation of ideas, languages, or popular culture through acculturation.

Definitions

According to the Oxford English Dictionary, the word 'globalization' was first employed in 1930, to denote a holistic view of human experience in education. An early description of globalization was penned by the American entrepreneur-turned-minister Charles Taze Russell who coined the term 'corporate giants' in 1897, although it was not until the 1960s that the term began to be widely used by economists and other social scientists. The term has since then achieved widespread use in the mainstream press by the later half of the 1980s. Since its inception, the concept of globalization has inspired numerous competing definitions and interpretations, with antecedents dating back to the great movements of trade and empire across Asia and the Indian Ocean from the 15th century onwards.

The United Nations ESCWA has written that globalization "is a widely-used term that can be defined in a number of different ways.

When used in an economic context, it refers to the reduction and removal of barriers between national borders in order to facilitate the flow of goods, capital, services and labour... although considerable barriers remain to the flow of labour... Globalization is not a new phenomenon. It began in the late nineteenth century, but it slowed down during the period from the start of the First World War until the third quarter of the twentieth century. This slowdown can be attributed to the inward-looking policies pursued by a number of countries in order to protect their respective industries... however, the pace of globalization picked up rapidly during the fourth quarter of the twentieth century..."

Saskia Sassen writes that "a good part of globalization consists of an enormous variety of micro-processes that begin to denationalize what had been constructed as national — whether policies, capital, political subjectivity, urban spaces, temporal frames, or any other of a variety of dynamics and domains."

Tom J. Palmer of the Cato Institute defines globalization as "the diminution or elimination of state-enforced restrictions on exchanges across borders and the increasingly integrated and complex global system of production and exchange that has emerged as a result."

Thomas L. Friedman has examined the impact of the "flattening" of the world, and argues that globalized trade, outsourcing, supply-chaining, and political forces have changed the world permanently, for both better and worse. He also argues that the pace of globalization is quickening and will continue to have a growing impact on business organization and practice.

Noam Chomsky argues that the word globalization is also used, in a doctrinal sense, to describe the neoliberal form of economic globalization.

Herman E. Daly argues that sometimes the terms internationalization and globalization are used interchangeably but there is a significant formal difference. The term "internationalization" (or internationalisation) refers to the importance of international trade, relations, treaties etc. owing to the (hypothetical) immobility of labour and capital between or among nations.

Finally, Takis Fotopoulos argues that globalization is the result of systemic trends manifesting the market economy's grow-or-die dynamic, following the rapid expansion of transnational corporations. Because these trends have not been offset effectively by counter-tendencies that could have emanated from trade-union action and other forms of political activity, the outcome has been globalisation. This is a multi-faceted and irreversible phenomenon within the system of the market economy and it is expressed as: economic globalisation, namely, the opening and deregulation of commodity, capital and labour markets which led to the present form of neoliberal globalisation; political globalisation, i.e., the emergence of a transnational elite and the phasing out of the all powerful-nation state of the statist period; cultural globalisation, i.e., the worldwide homogenisation of culture; ideological globalisation; technological globalisation; social globalisation.

History

The historical origins of globalization are the subject of on-going debate. Though some scholars situate the origins of globalization in the modern era, others regard it as a phenomenon with a long history. Perhaps the most extreme proponent of a deep historical origin for globalization was Andre Gunder Frank, an economist associated with dependency theory. Frank argued that a form of globalization has been in existence since the rise of trade links between Sumer and the Indus Valley Civilization in the third millennium B. C. Critics of this idea point out that it rests upon an over-broad definition of globalization.

An early form of globalized economics and culture existed during the Hellenistic Age, when commercialized urban centres were focused around the axis of Greek culture over a wide range that stretched from India to Spain, with such cities as Alexandria, Athens, and Antioch at its center. Trade was widespread during that period, and it is the first time the idea of a cosmopolitan culture (from Greek "Cosmopolis", meaning "world city") emerged. Others have perceived an early form of globalization in the trade links between the Roman Empire, the Parthian Empire, and the Han Dynasty. The increasing articulation of commercial links between these powers inspired the development of the Silk Road, which started in western China, reached the boundaries of the Parthian empire, and continued onwards towards Rome. With 300 Greek ships a year sailing between the Greco-Roman world and India, the annual trade may have reached 300,000 tons.

The Islamic Golden Age was also an important early stage of globalization, when Jewish and Muslim traders and explorers established a sustained economy across the Old World resulting in a globalization of crops, trade, knowledge and technology. Globally significant crops such as sugar and cotton became widely cultivated across the Muslim world in this period, while the necessity of learning Arabic and completing the Hajj created a cosmopolitan culture.

The advent of the Mongol Empire, though destabilizing to the commercial centres of the Middle East and China, greatly facilitated travel along the Silk Road. This permitted travellers and missionaries such as Marco Polo to journey successfully (and profitably) from one end of Eurasia to the other. The so-called Pax Mongolica of the thirteenth century had several other notable globalizing effects. It witnessed the creation of the first international postal service, as well as the rapid transmission of epidemic diseases such as bubonic plague across the newly unified regions of Central Asia. These pre-modern phases of global or hemispheric exchange are sometimes known as archaic globalization. Up to the sixteenth century, however, even the largest systems of international exchange were limited to the Old World.

The Age of Discovery brought a broad change in globalization, being the first period in which Eurasia and Africa engaged in substantial cultural, material and biologic exchange with the New World. It began in the late 15th century, when the two Kingdoms of the Iberian Peninsula-Portugal and Castile-sent the first exploratory voyages around the Horn of Africa and to the Americas, "discovered" in 1492 by Christopher Columbus. Shortly before the turn of the 16th century, Portuguese started establishing trading posts (factories) from Africa to Asia and Brazil, to deal with the trade of local products like gold, spices and timber, introducing an international business center under a royal monopoly, the House of India.

Global integration continued with the European colonization of the Americas initiating the Columbian Exchange, the enormous widespread exchange of plants, animals, foods, human populations (including slaves), communicable diseases, and culture between the Eastern and Western hemispheres. It was one of the most significant global events concerning ecology, agriculture, and culture in history. New crops that had come from the Americas via the European seafarers in the 16th century significantly contributed to the world's population growth.

This phase is sometimes known as proto-globalization. It was characterized by the rise of maritime European empires, in the 16th and 17th centuries, first the Portuguese and Spanish Empires, and later the Dutch and British Empires. In the 17th century, globalization became also a private business phenomenon when chartered companies like British East India Company (founded in 1600), often described as the first multinational corporation, as well as the Dutch East India Company (founded in 1602) were established. Because of the large investment and financing needs and high risks involved in international trade, the British East India Company became the first company in the world to share risk and enable joint ownership of companies through the issuance of shares of stock: an important driver for globalization.

The 19th century witnessed the advent of globalization approaching its modern form. Industrialization allowed cheap production of household items using economies of scale, while rapid population growth created sustained demand for commodities. Globalization in this period was decisively shaped by nineteenth-century imperialism. After the Opium Wars and the completion of British conquest of India, vast populations of these regions became ready consumers of European exports. It was in this period that areas of sub-Saharan Africa and the Pacific islands were incorporated into the world system. Meanwhile, the conquest of new parts of the globe, notably sub-Saharan Africa, by Europeans yielded valuable natural resources such as rubber, diamonds and coal and helped fuel trade and investment between the European imperial powers, their colonies, and the United States. Said John Maynard Keynes,

> *"The inhabitant of London could order by telephone, sipping his morning tea, the various products of the whole earth, and reasonably expect their early delivery upon his doorstep. Militarism and imperialism of racial and cultural rivalries were little more than the amusements of his daily newspaper. What an extraordinary episode in the economic progress of man was that age which came to an end in August 1914."*

The first phase of "modern globalization" began to break down at the beginning of the 20th century, with the first world war. The novelist VM Yeates criticised the financial forces of globalization as a factor in creating World War I. The final death knell for this phase came during the gold standard crisis and Great Depression in the late 1920s and early 1930s.

In the middle decades of the twentieth century globalization was largely driven by the global expansion of multinational corporations based in the United States and Europe, and worldwide exchange of new developments in science, technology and products, with most significant inventions of this time having their origins in the Western world according to Encyclopaedia Britannica. Worldwide export of western culture went through the new mass media: film, radio and television and recorded music. Development and growth of international transport and telecommunication played a decisive role in modern globalization.

In late 2000s, much of the industrialized world entered into a deep recession. Some analysts say the world is going through a period of deglobalization after years of increasing economic integration. Up to 45% of global wealth had been destroyed by the global financial crisis in little less than a year and a half. China has recently become the world's largest exporter surpassing Germany.

Modern Globalization

Globalization, since World War II, is largely the result of planning by politicians to break down borders hampering trade to increase prosperity and interdependence thereby decreasing the chance of future war. Their work led to the Bretton Woods conference, an agreement by the world's leading politicians to lay down the framework for international commerce and finance, and the founding of several international institutions intended to oversee the processes of globalization.

These institutions include the International Bank for Reconstruction and Development (the World Bank), and the International Monetary Fund. Globalization has been facilitated by advances in technology which have reduced the costs of trade, and trade negotiation rounds, originally under the auspices of the General Agreement on Tariffs and Trade (GATT), which led to a series of agreements to remove restrictions on free trade.

Since World War II, barriers to international trade have been considerably lowered through international agreements — GATT. Particular initiatives carried out as a result of GATT and the World Trade Organization (WTO), for which GATT is the foundation, have included:

- Promotion of free trade:
  - elimination of tariffs; creation of free trade zones with small or no tariffs
  - Reduced transportation costs, especially resulting from development of containerization for ocean shipping.
  - Reduction or elimination of capital controls
  - Reduction, elimination, or harmonization of subsidies for local businesses
  - Creation of subsidies for global corporations
  - Harmonization of intellectual property laws across the majority of states, with more restrictions

- o Supernational recognition of intellectual property restrictions (e.g. patents granted by China would be recognized in the United States).

Cultural globalization, driven by communication technology and the worldwide marketing of Western cultural industries, was understood at first as a process of homogenization, as the global domination of American culture at the expense of traditional diversity. However, a contrasting trend soon became evident in the emergence of movements protesting against globalization and giving new momentum to the defense of local uniqueness, individuality, and identity, but largely without success.

The Uruguay Round (1986 to 1994) led to a treaty to create the WTO to mediate trade disputes and set up a uniform platform of trading. Other bilateral and multilateral trade agreements, including sections of Europe's Maastricht Treaty and the North American Free Trade Agreement (NAFTA) have also been signed in pursuit of the goal of reducing tariffs and barriers to trade.

World exports rose from 8. 5% in 1970, to 16. 2% of total gross world product in 2001.

Measuring Globalization

Looking specifically at economic globalization demonstrates that it can be measured in different ways. These center around the four main economic flows that characterize globalization:

- Goods and services, e.g., exports plus imports as a proportion of national income or per capita of population
- Labour/people, e.g., net migration rates; inward or outward migration flows, weighted by population
- Capital, e.g., inward or outward direct investment as a proportion of national income or per head of population
- Technology, e.g., international research & development flows; proportion of populations (and rates of change thereof) using particular inventions (especially 'factor-neutral' technological advances such as the telephone, motorcar, broadband).

As globalization is not only an economic phenomenon, a multivariate approach to measuring globalization is the recent index calculated by the Swiss think tank KOF. The index measures the three main dimensions of globalization: economic, social, and political. In addition to three indices measuring these dimensions, an overall index of globalization and sub-indices referring to actual economic flows, economic restrictions, data on personal contact, data on information flows, and data on cultural proximity is calculated. Data is available on a yearly basis for 122 countries, as detailed in Dreher, Gaston and Martens (2008). According to the index, the world's most globalized country is Belgium, followed by Austria, Sweden, the United Kingdom and the Netherlands. The least globalized countries according to the KOF-index are Haiti, Myanmar, the Central African Republic and Burundi.

A. T. Kearney and *Foreign Policy Magazine* jointly publish another Globalization Index. According to the 2006 index, Singapore, Ireland, Switzerland, the Netherlands, Canada and Denmark are the most globalized, while Indonesia, India and Iran are the least globalized among countries listed.

Effects of Globalization

Globalization has various aspects which affect the world in several different ways such as:

- *Industrial*-emergence of worldwide production markets and broader access to a range of foreign products for consumers and companies. Particularly movement of material and goods between and within national boundaries. International trade in manufactured goods increased more than 100 times (from $95 billion to $12 trillion) in the 50 years since 1955. China's trade with Africa rose sevenfold during 2000-07 alone.
- *Financial*-emergence of worldwide financial markets and better access to external financing for borrowers. By the early part of the 21st century more than $1. 5 trillion in national currencies were traded daily to support the expanded levels of trade and investment. As these worldwide structures grew more quickly than any transnational regulatory regime, the instability of the global financial infrastructure dramatically increased, as evidenced by the Financial crisis of 2007–2010.
- *Economic*-realization of a global common market, based on the freedom of exchange of goods and capital. The interconnectedness of these markets, however, meant that an economic collapse in any one given country could not be contained.

Almost all notable worldwide IT companies are now present in India. Four Indians were among the world's top 10 richest in 2008, worth a combined $160 billion. In 2007, China had 415,000 millionaires and India 123,000.

- *Health Policy*-On the global scale, health becomes a commodity. In developing nations under the demands of Structural Adjustment Programs, health systems are fragmented and privatized. Global health policy makers have shifted during the 1990s from United Nations players to financial institutions. The result of this power transition is an increase in privatization in thé health sector. This privatization fragments health policy by crowding it with many players with many private interests. These fragmented policy players emphasize partnerships, specific interventions to combat specific problems (as opposed to comprehensive health strategies). Influenced by global trade and global economy, health policy is directed by technological advances and innovative medical trade. Global priorities, in this situation, are sometimes at odds with national priorities where increased health infrastructure and basic primary care are of more value to the public than privatized care for the wealthy.
- *Political*-some use "globalization" to mean the creation of a world government which regulates the relationships among governments and guarantees the rights arising from social and economic globalization. Politically, the United States has enjoyed a position of power among the world powers, in part because of its strong and wealthy economy. With the influence of globalization and with the help of The United States' own economy, the People's Republic of China has experienced some tremendous growth within the past decade. If China continues to grow at the rate projected by the trends, then it is very likely that in the next twenty years, there will be a major reallocation of power among the world leaders. China will have enough wealth, industry, and technology to rival the United States for the position of leading world power.
- *Informational*-increase in information flows between geographically remote locations. Arguably this is a technological change with the advent of fibre optic communications, satellites, and increased availability of telephone and Internet.
- *Language*-the most popular language is Mandarin (845 million speakers) followed by Spanish (329 million speakers) and English (328 million speakers).
  - About 35% of the world's mail, telexes, and cables are in English.
  - Approximately 40% of the world's radio programs are in English.
  - About 50% of all Internet traffic uses English.

- *Competition*-Survival in the new global business market calls for improved productivity and increased competition. Due to the market becoming worldwide, companies in various industries have to upgrade their products and use technology skillfully in order to face increased competition.
- *Ecological*-the advent of global environmental challenges that might be solved with international cooperation, such as climate change, cross-boundary water and air pollution, over-fishing of the ocean, and the spread of invasive species. Since many factories are built in developing countries with less environmental regulation, globalism and free trade may increase pollution. On the other hand, economic development historically required a "dirty" industrial stage, and it is argued that developing countries should not, via regulation, be prohibited from increasing their standard of living.
- *Cultural*-growth of cross-cultural contacts; advent of new categories of consciousness and identities which embodies cultural diffusion, the desire to increase one's standard of living and enjoy foreign products and ideas, adopt new technology and practices, and participate in a "world culture". Some bemoan the resulting consumerism and loss of languages. Also see Transformation of culture.
  - Spreading of multiculturalism, and better individual access to cultural diversity (e.g. through the export of Hollywood and, to a lesser extent, Bollywood movies). Some consider such "imported" culture a danger, since it may supplant the local culture, causing reduction in diversity or even assimilation. Others consider multiculturalism to promote peace and understanding between people.

    A third position gaining popularity is the notion that multiculturalism to a new form of monoculture in which no distinctions exist and everyone just shift between various lifestyles in terms of music, cloth and other aspects once more firmly attached to a single culture. Thus not mere cultural assimilation as mentioned above but the obliteration of culture as we know it today.
  - Greater international travel and tourism. WHO estimates that up to 500,000 people are on planes at any one time. In 2008, there were over 922 million international tourist arrivals, with a growth of 1. 9% as compared to 2007.

  - Greater immigration, including illegal immigration. The IOM estimates there are more than 200 million migrants around the world today. Newly available data show that remittance flows to developing countries reached $328 billion in 2008.
  - Spread of local consumer products (e.g., food) to other countries (often adapted to their culture).
  - Worldwide fads and pop culture such as Pokemon, Sudoku, Numa Numa, Origami, Idol series, YouTube, Orkut, Facebook, and MySpace. Accessible to those who have Internet or Television, leaving out a substantial segment of the Earth's population.
  - Worldwide sporting events such as FIFA World Cup and the Olympic Games.
  - Incorporation of multinational corporations into new media. As the sponsors of the All-Blacks rugby team, Adidas had created a parallel website with a downloadable interactive rugby game for its fans to play and compete.
- *Social*-development of the system of non-governmental organisations as main agents of global public policy, including humanitarian aid and developmental efforts.
- *Technical*
  - Development of a Global Information System, global telecommunications infrastructure and greater transborder data flow, using such technologies as the Internet, communication satellites, submarine fiber optic cable, and wireless telephones
  - Increase in the number of standards applied globally; e.g., copyright laws, patents and world trade agreements.
- *Legal/Ethical*
  - The creation of the international criminal court and international justice movements.
  - Crime importation and raising awareness of global crime-fighting efforts and cooperation.
  - The emergence of Global administrative law.
- *Religious*
  - The spread and increased interrelations of various religious groups, ideas, and practices and their ideas of the meanings and values of particular spaces.

Cultural Effects

"Culture" is defined as patterns of human activity and the symbols that give these activities significance. Culture is what people eat, how they dress, beliefs they hold, and activities they practice. Globalization has joined different cultures and made it into something different. As Erla Zwingle, from the National Geographic article titled "Globalization" states, "When cultures receive outside influences, they ignore some and adopt others, and then almost immediately start to transform them."

One classic culture aspect is food. Someone in America can be eating Japanese noodles for lunch while someone in Sydney, Australia is eating classic Italian meatballs. India is known for its curry and exotic spices. France is known for its cheeses. North America is known for its burgers and fries. McDonald's is a North American company which is now a global enterprise with 31,000 locations worldwide. This company is just one example of food causing cultural influence on the global scale.

Another common practice brought about by globalization is the usage of Chinese characters in tattoos. These tattoos are popular with today's youth despite the lack of social acceptance of tattoos in China. Also, there is a lack of comprehension in the meaning of Chinese characters that people get, making this an example of cultural appropriation.

The internet breaks down cultural boundaries across the world by enabling easy, near-instantaneous communication between people anywhere in a variety of digital forms and media. The Internet is associated with the process of cultural globalization because it allows interaction and communication between people with very different lifestyles and from very different cultures. Photo sharing websites allow interaction even where language would otherwise be a barrier.

Negative Effects

Globalization has been one of the most hotly debated topics in international economics over the past few years. Globalization has also generated significant international opposition over concerns that it has increased inequality and environmental degradation. In the Midwestern United States, globalization has eaten away at its competitive edge in industry and agriculture, lowering the quality of life in locations that lack the opportunity to adapt to the change.

Some also view the effect of globalization on culture as a rising concern. Along with globalization of economies and trade, culture is being imported and exported as well. The concern is that the stronger, bigger countries such as the United States, may overrun the other, smaller countries' cultures, leading to those customs and values being faded away. This process is also sometimes referred to as Americanization or McDonaldization.

Sweatshops

It can be said that globalization is the door that opens up an otherwise resource-poor country to the international market. Where a country has little material or physical product harvested or mined from its own soil, large corporations see an opportunity to take advantage of the "export poverty" of such a nation. Where the majority of the earliest occurrences of economic globalization are recorded as being the expansion of businesses and corporate growth, in many poorer nations globalization is actually the result of the foreign businesses investing in the country to take advantage of the lower wage rate: even though investing, by increasing the Capital Stock of the country, increases their wage rate.

One example used by anti-globalization protestors is the use of sweatshops by manufacturers. According to Global Exchange these "Sweat Shops" are widely used by sports shoe manufacturers and mentions one company in particular – Nike. There are factories set up in the poor countries where employees agree to work for low wages. Then if labour laws alter in those countries and stricter rules govern the manufacturing process the factories are closed down and relocated to other nations with more conservative, laissez-faire economic policies.

There are several agencies that have been set up worldwide specifically designed to focus on anti-sweatshop campaigns and education of such. In the USA, the National Labour Committee has proposed a number of bills as part of The Decent Working Conditions and Fair Competition Act, which have thus far failed in Congress. The legislation would legally require companies to respect human and worker rights by prohibiting the import, sale, or export of sweatshop goods.

Specifically, these core standards include no child labour, no forced labour, freedom of association, right to organize and bargain collectively, as well as the right to decent working conditions.

Tiziana Terranova has stated that globalization has brought a culture of "free labour". In a digital sense, it is where the individuals (contributing capital) exploits and eventually "exhausts the means through which labour can sustain itself". For example, in the area of digital media (animations, hosting chat rooms, designing games), where it is often less glamorous than it may sound. In the gaming industry, a Chinese Gold Market has been established.

Negative Effects of Economic Liberalization

The world today is so interconnected that the collapse of the subprime mortgage market in the U. S. has led to a global financial crisis and recession on a scale not seen since the Great Depression. Government deregulation and failed regulation of Wall Street's investment banks were important contributors to the subprime mortgage crisis. A flood of consumer goods such as televisions, radios, bicycles, and textiles into the United States, Europe, and Japan has helped fuel the economic expansion of Asian tiger economies in recent decades. However, Chinese textile and clothing exports have recently encountered criticism from Europe, the United States and some African countries. In South Africa, some 300,000 textile workers have lost their jobs due to the influx of Chinese goods. The increasing U. S. trade deficit with China has cost 2. 4 million American jobs between 2001 and 2008, according to a study by the Economic Policy Institute (EPI). A total of 3. 2 million – one in six U. S. factory jobs – have disappeared between 2000 and 2007.

Brain Drain

Opportunities in richer countries drives talent away from poorer countries, leading to brain drains. Brain drain has cost the African continent over $4. 1 billion in the employment of 150,000 expatriate professionals annually. Indian students going abroad for their higher studies costs India a foreign exchange outflow of $10 billion annually. Rowden (2009) argues in "*The Deadly Ideas of Neoliberalism: How the IMF has Undermined Public Health and the Fight Against Aids*" (London: Zed Books), that the International Monetary Fund's monetarist approach towards prioritizing price stability (low inflation) and fiscal restraint (low budget deficits) was unnecessarily restrictive and has prevented developing countries from being able to increase long-term public investment as a percent of GDP in the underlying public health infrastructure.

Rowden claims the consequences of these and other neoliberal policies have been chronically underfunded public health systems, dilapidated health infrastructures, inadequate numbers of health personnel, and demoralizing working conditions that have fuelled the "push factors" driving the brain drain of doctors and nurses migrating from poor countries to rich ones, all of which has undermined public health systems and the fight against HIV/AIDS in developing countries.

Effect on Environmental Degradation

The Worldwatch Institute said the booming economies of China and India are planetary powers that are shaping the global biosphere. In 2007, China overtook the United States as the world's biggest producer of CO_2. At present rates, tropical rainforests in Indonesia would be logged out in 10 years, Papua New Guinea in 13 to 16 years. A major source of deforestation is the logging industry, driven spectacularly by China and Japan. Thriving economies such as China and India are quickly becoming large oil consumers. China has seen oil consumption grow by 8% yearly since 2002, doubling from 1996–2006. Crude oil prices in the last several years have steadily risen from about $25 a barrel in August 2003 to over $140 a barrel in July 2008. *State of the World* 2006 report said the two countries' high economic growth hid a reality of severe pollution. The report states:

> *The world's ecological capacity is simply insufficient to satisfy the ambitions of China, India, Japan, Europe and the United States as well as the aspirations of the rest of the world in a sustainable way.*

Without more recycling, zinc could be used up by 2037, both indium and hafnium could run out by 2017, and terbium could be gone before 2012. It is said that if China and India were to consume as much resources per capita as United States or Japan in 2030 together they would require a full planet Earth to meet their needs. In the longterm these effects can lead to increased conflict over dwindling resources and in the worst case a Malthusian catastrophe.

Food Security

The head of the International Food Policy Research Institute, stated in 2008 that the gradual change in diet among newly prosperous populations is the most important factor underpinning the rise in global food prices. From 1950 to 1984, as the Green Revolution transformed agriculture around the world, grain production increased by over 250%. The world population has grown by about 4 billion since the beginning of the Green Revolution and most believe that, without the Revolution, there would be greater famine and malnutrition than the UN presently documents (approximately 850 million people suffering from chronic malnutrition in 2005).

It is becoming increasingly difficult to maintain food security in a world beset by a confluence of "peak" phenomena, namely peak oil, peak water, peak phosphorus, peak grain and peak fish. Growing populations, falling energy sources and food shortages will create the "perfect storm" by 2030, according to the UK government chief scientist. He said food reserves are at a 50-year low but the world requires 50% more energy, food and water by 2030.

The world will have to produce 70% more food by 2050 to feed a projected extra 2. 3 billion people and as incomes rise, the United Nations' Food and Agriculture Organisation (FAO) warned. Social scientists have warned of the possibility that global civilization is due for a period of contraction and economic re-localization, due to the decline in fossil fuels and resulting crisis in transportation and food production. One paper even suggested that the future might even bring about a restoration of sustainable local economic activities based on hunting and gathering, shifting horticulture, and pastoralism. The journal *Science* published a four-year study in November 2006, which predicted that, at prevailing trends, the world would run out of wild-caught seafood in 2048.

Effect on Disease

Globalization, the flow of information, goods, capital and people across political and geographic boundaries, has also helped to spread some of the deadliest infectious diseases known to humans. Starting in Asia, the Black Death killed at least one-third of Europe's population in the 14th century. Even worse devastation was inflicted on the American supercontinent by Europe. For instance 90% of the populations of the civilizations of the "New World" such as the Aztec, Maya, and Inca were killed by small pox brought by European colonization. Modern modes of transportation allow more people and products to travel around the world at a faster pace, they also open the airways to the transcontinental movement of infectious disease vectors. One example of this occurring is AIDS/HIV. Approximately 1. 1 million persons are living with HIV/AIDS in the United States, and AIDS remains the leading cause of death among African American women between ages 25 and 34. Due to immigration, approximately 500,000 people in the United States are believed to be infected with Chagas disease. In 2006, the tuberculosis (TB) rate among foreign-born persons in the United States was 9. 5 times that of U. S.-born persons.

Drug and Illicit Goods Trade

The United Nations Office on Drugs and Crime (UNODC) issued a report that the global drug trade generates more than $320 billion a year in revenues. Worldwide, the UN estimates there are more than 50 million regular users of heroin, cocaine and synthetic drugs. The international trade of endangered species is second only to drug trafficking. Traditional Chinese medicine often incorporates ingredients from all parts of plants, the leaf, stem, flower, root, and also ingredients from animals and minerals. The use of parts of endangered species (such as seahorses, rhinoceros horns, saiga antelope horns, and tiger bones and claws) has created controversy and resulted in a black market of poachers who hunt restricted animals. In 2003, 29% of open sea fisheries were in a state of collapse.

Pro-globalization (Globalism)

Supporters of free trade claim that it increases economic prosperity as well as opportunity, especially among developing nations, enhances civil liberties and leads to a more efficient allocation of resources. Economic theories of comparative advantage suggest that free trade leads to a more efficient allocation of resources, with all countries involved in the trade benefiting. In general, this leads to lower prices, more employment, higher output and a higher standard of living for those in developing countries.

Dr. Francesco Stipo, Director of the USA Club of Rome suggests that "the world government should reflect the political and economic balances of world nations. A world confederation would not supersede the authority of the State governments but rather complement it, as both the States and the world authority would have power within their sphere of competence".

Proponents of laissez-faire capitalism, and some libertarians, say that higher degrees of political and economic freedom in the form of democracy and capitalism in the developed world are ends in themselves and also produce higher levels of material wealth. They see globalization as the beneficial spread of liberty and capitalism.

Supporters of democratic globalization are sometimes called pro-globalists. They believe that the first phase of globalization, which was market-oriented, should be followed by a phase of building global political institutions representing the will of world citizens. The difference from other globalists is that they do not define in advance any ideology to orient this will, but would leave it to the free choice of those citizens via a democratic process. Some, such as former Canadian Senator Douglas Roche, O. C., simply view globalization as inevitable and advocate creating institutions such as a directly elected United Nations Parliamentary Assembly to exercise oversight over unelected international bodies.

Anti-globalization

The "anti-globalization movement" is a term used to describe the political group who oppose the neoliberal version of globalization, while criticisms of globalization are some of the reasons used to justify this group's stance.

"Anti-globalization" may also involve the process or actions taken by a state or its people in order to demonstrate its sovereignty and practice democratic decision-making. Anti-globalization may occur in order to maintain barriers to the international transfer of people, goods and beliefs, particularly free market deregulation, encouraged by organizations such as the International Monetary Fund or the World Trade Organization. Moreover, as Naomi Klein argues in her book No Logo, anti-globalism can denote either a single social movement or an umbrella term that encompasses a number of separate social movements such as nationalists and socialists.

In either case, participants stand in opposition to the unregulated political power of large, multi-national corporations, as the corporations exercise power through leveraging trade agreements which in some instances damage the democratic rights of citizens, the environment particularly air quality index and rain forests, as well as national government's sovereignty to determine labour rights, including the right to form a union, and health and safety legislation, or laws as they may otherwise infringe on cultural practices and traditions of developing countries.

Some people who are labelled "anti-globalist" or "sceptics" (Hirst and Thompson) consider the term to be too vague and inaccurate. Podobnik states that "the vast majority of groups that participate in these protests draw on international networks of support, and they generally call for forms of globalization that enhance democratic representation, human rights, and egalitarianism."

Joseph Stiglitz and Andrew Charlton write:

> *"The anti-globalization movement developed in opposition to the perceived negative aspects of globalization. The term 'anti-globalization' is in many ways a misnomer, since the group represents a wide range of interests and issues and many of the people involved in the anti-globalization movement do support closer ties between the various peoples and cultures of the world through, for example, aid, assistance for refugees, and global environmental issues."*

Some members aligned with this viewpoint prefer instead to describe themselves as the "Global Justice Movement", the "Anti-Corporate-Globalization Movement", the "Movement of Movements" (a popular term in Italy), the "Alter-globalization" movement (popular in France), the "Counter-Globalization" movement, and a number of other terms. Critiques of the current wave of economic globalization typically look at both the damage to the planet, in terms of the perceived unsustainable harm done to the biosphere, as well as the perceived human costs, such as poverty, inequality, miscegenation, injustice and the erosion of traditional culture which, the critics contend, all occur as a result of the economic transformations related to globalization. They challenge directly the metrics, such as GDP, used to measure progress promulgated by institutions such as the World Bank, and look to other measures, such as the Happy Planet Index, created by the New Economics Foundation. They point to a "multitude of interconnected fatal consequences–social disintegration, a breakdown of democracy, more rapid and extensive deterioration of the environment, the spread of new diseases, increasing poverty and alienation" which they claim are the unintended but very real consequences of globalization.

The terms globalization and anti-globalization are used in various ways. Noam Chomsky believes that,

> *"The term "globalization" has been appropriated by the powerful to refer to a specific form of international economic integration, one based on investor rights, with the interests of people incidental. That is why the business press, in its more honest moments, refers to the "free trade agreements" as "free investment agreements" (Wall St. Journal). Accordingly, advocates of other forms of globalization are described as "anti-globalization"; and some, unfortunately, even accept this term, though it is a term of propaganda that should be dismissed with ridicule. No sane person is opposed to globalization, that is, international integration. Surely not the left and the workers movements, which were founded on the principle of international solidarity — that is, globalization in a form that attends to the rights of people, not private power systems."*
>
> *"The dominant propaganda systems have appropriated the term "globalization" to refer to the specific version of international economic integration that they favour,*

> *which privileges the rights of investors and lenders, those of people being incidental. In accord with this usage, those who favour a different form of international integration, which privileges the rights of human beings, become "anti-globalist." This is simply vulgar propaganda, like the term "anti-Soviet" used by the most disgusting commissars to refer to dissidents. It is not only vulgar, but idiotic. Take the World Social Forum, called "anti-globalization" in the propaganda system – which happens to include the media, the educated classes, etc., with rare exceptions. The WSF is a paradigm example of globalization. It is a gathering of huge numbers of people from all over the world, from just about every corner of life one can think of, apart from the extremely narrow highly privileged elites who meet at the competing World Economic Forum, and are called "pro-globalization" by the propaganda system. An observer watching this farce from Mars would collapse in hysterical laughter at the antics of the educated classes."*

Critics argue that globalization results in:

Poorer countries suffering disadvantages: While it is true that globalization encourages free trade among countries, there are also negative consequences because some countries try to save their national markets. The main export of poorer countries is usually agricultural goods. Larger countries often subsidise their farmers (like the EU Common Agricultural Policy), which lowers the market price for the poor farmer's crops compared to what it would be under free trade.

The exploitation of foreign impoverished workers: The deterioration of protections for weaker nations by stronger industrialized powers has resulted in the exploitation of the people in those nations to become cheap labour. Due to the lack of protections, companies from powerful industrialized nations are able to offer workers enough salary to entice them to endure extremely long hours and unsafe working conditions, though economists question if consenting workers in a competitive employers' market can be decried as "exploited". It is true that the workers are free to leave their jobs, but in many poorer countries, this would mean starvation for the worker, and possible even his/her family if their previous jobs were unavailable.

The shift to outsourcing: The low cost of offshore workers have enticed corporations to buy goods and services from foreign countries.

The laid off manufacturing sector workers are forced into the service sector where wages and benefits are low, but turnover is high. This has contributed to the deterioration of the middle class which is a major factor in the increasing economic inequality in the United States. Families that were once part of the middle class are forced into lower positions by massive layoffs and outsourcing to another country. This also means that people in the lower class have a much harder time climbing out of poverty because of the absence of the middle class as a stepping stone.

Weak labour unions: The surplus in cheap labour coupled with an ever growing number of companies in transition has caused a weakening of labour unions in the United States. Unions lose their effectiveness when their membership begins to decline. As a result unions hold less power over corporations that are able to easily replace workers, often for lower wages, and have the option to not offer unionized jobs anymore.

An increase in exploitation of child labour: for example, a country that experiencing increases in labour demand because of globalization and an increase the demand for goods produced by children, will experience greater a demand for child labour. This can be "hazardous" or "exploitive", e.g., quarrying, salvage, cash cropping but also includes the trafficking of children, children in bondage or forced labour, prostitution, pornography and other illicit activities.

In December 2007, World Bank economist Branko Milanovic has called much previous empirical research on global poverty and inequality into question because, according to him, improved estimates of purchasing power parity indicate that developing countries are worse off than previously believed. Milanovic remarks that "literally hundreds of scholarly papers on convergence or divergence of countries' incomes have been published in the last decade based on what we know now were faulty numbers." With the new data, possibly economists will revise calculations, and he also believed that there are considerable implications estimates of global inequality and poverty levels. Global inequality was estimated at around 65 Gini points, whereas the new numbers indicate global inequality to be at 70 on the Gini scale.

The critics of globalization typically emphasize that globalization is a process that is mediated according to corporate interests, and typically raise the possibility of alternative global institutions and policies, which they believe address the moral claims of poor and working classes throughout the globe, as well as environmental concerns in a more equitable way.

The movement is very broad, including church groups, national liberation factions, peasant unionists, intellectuals, artists, protectionists, anarchists, those in support of relocalization and others. Some are reformist, (arguing for a more moderate form of capitalism) while others are more revolutionary (arguing for what they believe is a more humane system than capitalism) and others are reactionary, believing globalization destroys national industry and jobs.

One of the key points made by critics of recent economic globalization is that income inequality, both between and within nations, is increasing as a result of these processes. One article from 2001 found that significantly, in 7 out of 8 metrics, income inequality has increased in the twenty years ending 2001. Also, "incomes in the lower deciles of world income distribution have probably fallen absolutely since the 1980s". Furthermore, the World Bank's figures on absolute poverty were challenged. The article was skeptical of the World Bank's claim that the number of people living on less than $1 a day has held steady at 1. 2 billion from 1987 to 1998, because of biased methodology.

A chart that gave the inequality a very visible and comprehensible form, the so-called 'champagne glass' effect, was contained in the 1992 United Nations Development Program Report, which showed the distribution of global income to be very uneven, with the richest 20% of the world's population controlling 82. 7% of the world's income.

***Table:** Distribution of World GDP, 1989*

Quintile of Population	***Income***
Richest 20%	82. 7%
Second 20%	11. 7%
Third 20%	2. 3%
Fourth 20%	2. 4%
Poorest 20%	0. 2%

Source: United Nations Development Program. 1992 Human Development Report

Economic arguments by fair trade theorists claim that unrestricted free trade benefits those with more financial leverage (i.e. the rich) at the expense of the poor.

Americanization related to a period of high political American clout and of significant growth of America's shops, markets and object being brought into other countries. So globalization, a much more diversified phenomenon, relates to a multilateral political world and to the increase of objects, markets and so on into each others countries.

Critics of globalization talk of Westernization. A 2005 UNESCO report showed that cultural exchange is becoming more frequent from Eastern Asia but Western countries are still the main exporters of cultural goods. In 2002, China was the third largest exporter of cultural goods, after the UK and US. Between 1994 and 2002, both North America's and the European Union's shares of cultural exports declined, while Asia's cultural exports grew to surpass North America. Related factors are the fact that Asia's population and area are several times that of North America. Some opponents of globalization see the phenomenon as the promotion of corporatist interests. They also claim that the increasing autonomy and strength of corporate entities shapes the political policy of countries.

International Social Forums

The first WSF in 2001 was an initiative of the administration of Porto Alegre in Brazil. The slogan of the World Social Forum was "Another World Is Possible". It was here that the WSF's Charter of Principles was adopted to provide a framework for the forums. It proved to be a vibrant player in macro level.

The WSF became a periodic meeting: in 2002 and 2003 it was held again in Porto Alegre and became a rallying point for worldwide protest against the American invasion of Iraq. In 2004 it was moved to Mumbai (formerly known as Bombay, in India), to make it more accessible to the populations of Asia and Africa. This last appointment saw the participation of 75,000 delegates.

In the meantime, regional forums took place following the example of the WSF, adopting its Charter of Principles. The first European Social Forum (ESF) was held in November 2002 in Florence. The slogan was "Against the war, against racism and against neo-liberalism". It saw the participation of 60,000 delegates and ended with a huge demonstration against the war (1,000,000 people according to the organizers). The other two ESFs took place in Paris and London, in 2003 and 2004 respectively.

Recently there has been some discussion behind the movement about the role of the social forums. Some see them as a "popular university", an occasion to make many people aware of the problems of globalization. Others would prefer that delegates concentrate their efforts on the coordination and organization of the movement and on the planning of new campaigns. However it has often been argued that in the dominated countries (most of the world) the WSF is little more than an 'NGO fair' driven by Northern NGOs and donors most of which are hostile to popular movements of the poor.

Relation to Americanization

In the past, the argument that globalization could in actuality have been the spreading of American culture could have been made. The thorough spreading of American culture throughout the world was apparent. For example, there were cola products shipped and sold in nearly every country in the world. American fashion also seemed to be trending in most other countries. Alongside this, the biggest economy and strongest army in the U. S. A could only help the impact of their assimilation of the rest of the world. Many of their inventions, like the iPod, have made their way around the world, and certainly, especially the technology, will remain for a long period of time. Their domination in so many world sports and creation of many others only provide more influence.

However, despite this, in current times (2010) there can be made a new argument. America has been knocked off its pedestal of the greatest world power, and is making way for new ones, namely China and India. The spreading of Chinese products throughout the world is occurring rapidly. They have an, arguably, unmatched economy in the world. In addition to this, India is quickly becoming well known for its products too. The large country is the creator and supplier for mostly clothing, but other products too, for many countries in the world. These two countries are rapidly replacing American culture as the most widespread example of globalization in the world.

Chapter 5

Events of Economic Crisis

2007–2008 World Food Price Crisis

The years 2007–2008 saw dramatic increases in world food prices, creating a global crisis and causing political and economical instability and social unrest in both poor and developed nations.

Systemic causes for the worldwide increases in food prices continue to be the subject of debate. Initial causes of the late 2006 price spikes included droughts in grain-producing nations and rising oil prices. Oil price increases also caused general escalations in the costs of fertilizers, food transportation, and industrial agriculture.

Root causes may be the increasing use of biofuels in developed countries, and an increasing demand for a more varied diet across the expanding middle-class populations of Asia.

These factors, coupled with falling world-food stockpiles all contributed to the worldwide rise in food prices.

Causes not commonly attributed by mainstream views include structural changes in trade and agricultural production, agricultural price supports and subsidies in developed nations, diversions of food commodities to high input foods and fuel, commodity market speculation, and climate change.

Drastic Price Increases

Between 2006 and 2008 average world prices for rice rose by 217%, wheat by 136%, corn by 125% and soybeans by 107%. In late April 2008 rice prices hit 24 cents (U. S.) per U. S. pound, more than doubling the price in just seven months.

World Population Growth

Although some commentators have argued that this food crisis stems from unprecedented global population growth, others point out that world population growth rates have dropped dramatically since the 1980s, and grain availability has continued to outpace population. However, despite production gains made in the last decade, world food demand outpaces any production increases. According to Joachim von Braun, of the IFPRI, total food production increases only about 1 to 2 percent per year, while total world population increases approximately 4%. Aggregate cereal grain food production, per capita, had risen yearly from the 1960s to the 1980s but has been in decline since. However, this does not take into account any non-food uses of grain production. World population has grown from 1. 6 billion in 1900 to an estimated 6. 8 billion.

Increased Demand for More Resource Intensive Food

The head of the International Food Policy Research Institute, stated in 2008 that the gradual change in diet among newly prosperous populations is the most important factor underpinning the rise in global food prices. Where food utilization has increased, it has largely been in processed ("value added") foods, sold in developing and developed nations. Total grain utilization growth since 2006 (up three percent, over the 2000–2006 per annum average of two percent) has been greatest in non-food usage, especially in feed and biofuels. One kilogram of beef requires seven kilograms of feed grain. These reports, therefore, conclude that usage in industrial, feed, and input intensive foods, not population growth among poor consumers of simple grains, has contributed to the price increases.

Table: *2005/1990 Ratios of Per Capita Consumption*

	India	China	Brazil	Nigeria
Cereals	1. 0	0. 8	1. 2	1. 0
Meat	1. 2	2. 4	1. 7	1. 0
Milk	1. 2	3. 0	1. 2	1. 3
Fish	1. 2	2. 3	0. 9	0. 8
Fruits	1. 3	3. 5	0. 8	1. 1
Vegetables	1. 3	2. 9	1. 3	1. 3

Although the vast majority of the population in Asia remains rural and poor, the growth of the middle class in the region has been dramatic. For comparison, in 1990, the middle class grew by 9. 7 percent in India and 8. 6 percent in China, but by 2007 the growth rate was nearly 30 percent and 70 percent respectively. The corresponding increase in Asian affluence also brought with it a change in lifestyle and eating habits, particularly a demand for greater variety, leading to increased competition with western nations for already strained agricultural resources. This demand exacerbates dramatic increases in commodity prices such as oil. Another issue was rising affluence in India and China was reducing the 'shock absorber' of poor people who are forced to reduce their resource consumption when food prices rise. This reduced price elasticity and caused a sharp rise in food prices during some shortages. In the media, China is often mentioned as one of the main reasons for the increase in world food prices. However, China has to a large extent been able to meet its own demand for food, and even exports its surpluses in the world market.

Impact of Petroleum Price Increases

The rise in the price of oil has heightened the costs of fertilizers (in some instances doubling the price within the six months before April, 2008), the majority of which require petroleum or natural gas to manufacture. Although the main fossil fuel input for fertilizer comes from natural gas to generate hydrogen for the Haber–Bosch process, natural gas has its own supply problems similar to those for oil. Because natural gas can substitute for petroleum in some uses (for example, natural gas liquids and electricity generation), increasing prices for petroleum lead to increasing prices for natural gas, and thus for fertilizer. Costs for fertilizer raw materials other than oil, such as potash, have themselves been increasing as increased production of staples increases demand. This is causing a boom (with associated volatility) in agriculture stocks.

Declining World Food Stockpiles

In the past, nations tended to keep more sizable food stockpiles, but more recently, due to a faster pace of food growth and ease of importation, less emphasis is placed on high stockpiles. For example, in February 2008 wheat stockpiles hit a 60-year low in the United States. Data stocks are often calculated as a residual between Production and Consumption but it becomes difficult to discriminate between a destocking policy choices of individual countries and a deficit between production and consumption.

Financial Speculation

Destabilizing influences, including indiscriminate lending and real estate speculation, led to a crisis in January 2008, and eroded investment in food commodities. The United States, specifically, had been facing an economic crisis that eventually lead to recession. Financial speculation in commodity futures following the collapse of the financial derivatives markets has contributed to the crisis due to a "commodities super-cycle." Financial speculators seeking quick returns have removed trillions of dollars from equities and mortgage bonds, some of which has been invested into food and raw materials. That American commodities speculation could have a worldwide impact on food prices is reflected in the globalization of food production. It represents the concentration of wealth throughout the world, which Frances Moore Lappe equates to a weakening in fundamental democracy. In a recent article for *The Nation*, she suggests that there is no food shortage but that "as long as food is merely a commodity in societies that don't protect people's right to participate in the market, and as long as farming is left vulnerable to consolidated power off the farm, many will go hungry, farmers among them—no matter how big the harvests."

Commodity Index Funds

A 2010 article in Harper's by Frederick Kaufman alleged that these Commodity Index funds (invented by Goldman Sachs) were part of the global food crisis in 2008, including riots in 30 countries, a food price rise of 80 percent between 2005 and 2008, and an increased hunger rate. The article claimed that one mechanism involved was a 'demand shock' on wheat futures caused by the index funds, resulting in a 'contango' wheat market on the Chicago Mercantile Exchange. This all caused prices of wheat to rise much higher than normal, defeating the purpose of the exchanges (price stabilization) in the first place.

Impact of Trade Liberalization

Some theorists, such as Martin Khor of the Third World Network, point out that many developing nations have gone from being food independent to being net food importing economies since the 1970s and 1980s International Monetary Fund (and later the World Trade Organisation's Agreement on Agriculture) free market economics directives to debtor nations. In opening developing countries to developed world food imports subsidised by Western governments, developing nations have become dependent upon food imports that are cheaper than what local smallholders agriculture produces, even in the poorest regions of the world.

While developed countries pressured the developing world to abolish subsidies in the interest of trade liberalization, rich countries largely kept subsidies in place for their own farmers. In recent years

United States government subsidies have been added to push production toward biofuel rather than food and vegetables.

Impact of Food for Fuel

One systemic cause for the price rise is held to be the diversion of food crops (maize in particular) for making first-generation biofuels. An estimated 100 million tons of grain per year are being redirected from food to fuel. (Total worldwide grain production for 2007 was just over 2000 million tonnes.) As farmers devoted larger parts of their crops to fuel production than in previous years, land and resources available for food production were reduced correspondingly. This has resulted in less food available for human consumption, especially in developing and least developed countries, where a family's daily allowances for food purchases are extremely limited. The crisis can be seen, in a sense, to dichotomize rich and poor nations, since, for example, filling a tank of an average car with biofuel, amounts to as much maize (Africa's principal food staple) as an African person consumes in an entire year.

Brazil, the world's second largest producer of ethanol after the U. S., is considered to have the world's first sustainable biofuels economy and its government claims Brazil's sugar cane based ethanol industry has not contributed to the 2008 food crises. A World Bank policy research working paper released in July 2008 concluded that "... large increases in biofuels production in the United States and Europe are the main reason behind the steep rise in global food prices", and also stated that "Brazil's sugar-based ethanol did not push food prices appreciably higher". An economic assessment published in July 2008 by the OECD disagrees with the World Bank report regarding the negative effects of subsidies and trade restrictions, finding that the impact of biofuels on food prices are much smaller.

A report released by Oxfam in June 2008 criticized biofuel policies of rich countries and concluded that from all biofuels available in the market, Brazilian sugarcane ethanol is "far from perfect" but it is the most favourable biofuel in the world in term of cost and GHG balance. The report discusses some existing problems and potential risks, and asks the Brazilian government for caution to avoid jeopardazing its environmental and social sustainability. The report also says that: "Rich countries spent up to $15 billion last year supporting biofuels while blocking cheaper Brazilian ethanol, which is far less damaging for global food security."

German Chancellor Angela Merkel said the rise in food prices is due to poor agricultural policies and changing eating habits in developing nations, not biofuels as some critics claim. On April 29, 2008, U. S. President George W. Bush declared during a press conference that "85 percent of the world's food prices are caused by weather, increased demand and energy prices", and recognized that "15 percent has been caused by ethanol". On July 4, 2008, The Guardian reported that a leaked World Bank report estimated the rise in food prices caused by biofuels to be 75%. This report was officially released in July 2008.

Since reaching record high prices in June 2008, corn prices fell 50% by October 2008, declining sharply together with other commodities, including oil. As ethanol production from corn has continue at the same levels, some have argued this trend shows the belief that the increased demand for corn to produce ethanol was mistaken. "Analysts, including some in the ethanol sector, say ethanol demand adds about 75 cents to $1. 00 per bushel to the price of corn, as a rule of thumb. Other analysts say it adds around 20 percent, or just under 80 cents per bushel at current prices. Those estimates hint that $4 per bushel corn might be priced at only $3 without demand for ethanol fuel." These industry sources consider that a speculative bubble in the commodity markets holding positions in corn futures was the main driver behind the observed hike in corn prices affecting food supply.

Second-and third-generation biofuels (such as cellulosic ethanol and algae fuel, respectively) may someday ease the competition with food crops, as non food energy crops can grow on marginal lands unsuited for food crops, but these advanced biofuels require further development of farming practices and refining technology; in contrast, ethanol from maize uses mature technology and the maize crop can be shifted between food and fuel use quickly.

Biofuel Subsidies in the US and the EU

The World Bank lists the effect of biofuels as an important contributor to higher food prices. The FAO/ECMB has reported that world land usage for agriculture has declined since the 1980s, and subsidies outside the United States and EU have dropped since the year 2004, leaving supply, while sufficient to meet 2004 needs, vulnerable when the United States began converting agricultural commodities to biofuels. According to the United States Department of Agriculture, global wheat imports and stocks have decreased, domestic consumption has stagnated, and world wheat production has decreased from 2006 to 2008.

In the United States, government subsidies for ethanol production have prompted many farmers to switch to production for biofuel. Maize is the primary crop used for the production of ethanol, with the United

States being the biggest producer of maize ethanol. As a result, 23 percent of United States maize crops were being used for ethanol in 2006–2007 (up from 6 percent in 2005–2006), and the USDA expects the United States to use 81 million tonnes of maize for ethanol production in the 2007–2008 season, up 37 percent. This not only diverts grains from food, but it diverts agricultural land from food production.

Nevertheless, supporters of ethanol claim that using corn for ethanol is not responsible for the worst food riots in the world, many of which have been caused by the price of rice and oil, which are not affected by biofuel use but rather by supply and demand.

However, a World Bank policy research working paper released in July 2008 says that biofuels have raised food prices between 70 to 75 percent. The study found that higher oil prices and a weak dollar explain 25–30% of total price rise. The "month-by-month" five year analysis disputes that increases in global grain consumption and droughts were responsible for price increases, reporting that this had only a marginal impact and instead argues that the EU and US drive for biofuels has had by far the biggest impact on food supply and prices. The paper concludes that increased production of biofuels in the US and EU were supported by subsidies and tariffs on imports, and considers that without these policies, price increases would have been smaller. This research also concluded that Brazil's sugar cane based ethanol has not raised sugar prices significantly, and suggest to remove tariffs on ethanol imports by both the US and EU, to allow more efficient producers such as Brazil and other developing countries to produce ethanol profitably for export to meet the mandates in the EU and the US. The Renewable Fuel Association (RFA) published a rebuttal based on the version leaked before the formal release of the World Bank's paper. The RFA critique considers that the analysis is highly subjective and that the author "*estimates the impact of global food prices from the weak dollar and the direct and indirect effect of high petroleum prices and attributes everything else to biofuels.*"

An economic assessment report also published in July 2008 by the OECD agrees with the World Bank report regarding the negative effects of subsidies and trade restrictions, but found that the impact of biofuels on food prices are much smaller. The OECD study is also critical of the limited reduction of GHG emissions achieved from biofuels produced in Europe and North America, concluding that the current biofuel support policies would reduce greenhouse gas emissions from transport fuel by no more than 0. 8 percent by 2015, while Brazilian ethanol from sugar cane reduces greenhouse gas emissions by at least 80 percent compared to fossil fuels. The assessment calls on governments for more open markets in biofuels and feedstocks to improve efficiency and lower costs.

Idled Farmland

According to the *New York Times* on April 9, 2008, the United States government pays farmers to idle their cropland under a conservation program. This policy reached a peak of 36,800,000 acres (149,000 km^2) idled in 2007, that is 8% of cropland in United States, representing a total area bigger than the state of New York.

Agricultural Subsidies

The global food crisis has renewed calls for removal of distorting agricultural subsidies in developed countries. Support to farmers in OECD countries totals 280 billion USD annually, which compares to official development assistance of just 80 billion USD in 2004, and farm support distorts food prices leading to higher global food prices, according to OECD estimates. The US Farm Bill brought in by the Bush Administration in 2002 increased agricultural subsidies by 80% and cost the US taxpayer 190 billion USD.. In 2003, the EU agreed to extend the Common Agricultural Policy until 2013. Former UNDP Administrator Malloch Brown renewed calls for reform of the farm subsidies such as the CAP.

Distorted Global Rice Market

Japan is forced to import more than 767,000 tons of rice annually from the United States, Thailand, and other countries due to WTO rules. This is despite the fact that Japan produces over 100% of domestic rice consumption needs with 11 million tonnes produced in 2005 while 8. 7 million tonnes were consumed in 2003–2004 period. Japan is not allowed to re-export this rice to other countries without approval. This rice is generally left to rot and then used for animal feed. Under pressure, the United States and Japan are poised to strike a deal to remove such restrictions. It is expected 1. 5 million tonnes of high-grade American rice will enter the market soon.

Crop Shortfalls from Natural Disasters

Several distinct weather-and climate-related incidents have caused disruptions in crop production. Perhaps the most influential is the extended drought in Australia, in particular the fertile Murray-Darling Basin, which produces large amounts of wheat and rice. The drought has caused the annual rice harvest to fall by as much as 98% from pre-drought levels. Australia is historically the second-largest exporter of wheat after the United States, producing up to 25 million tons in a good year, the vast majority for export. However, the 2006 harvest was 9. 8 million. Other events that have negatively affected the price of food include the 2006 heat wave in California's San Joaquin Valley, which killed large numbers of farm animals, and unseasonable 2008 rains in Kerala, India, which destroyed swathes of grain. Scientists have stated that several of these incidents are consistent with the predicted effects of climate change.

The effects of Cyclone Nargis on Burma in May 2008 caused a spike in the price of rice. Burma has historically been a rice exporter, though yields have fallen as government price controls have reduced incentives for farmers. The storm surge inundated rice paddies up to 30 miles (48 km) inland in the Irrawaddy Delta, raising concern that the salt could make the fields infertile. The FAO had previously estimated that Burma would export up to 600,000 tons of rice in 2008, but concerns were raised in the cyclone's aftermath that Burma may be forced to import rice for the first time, putting further upward pressure on global rice prices.

Stem rust reappeared in 1998 in Uganda (and possibly earlier in Kenya) with the particularly virulent UG99 fungus. Unlike other rusts, which only partially affect crop yields, UG99 can bring 100% crop loss. Up to 80% yield losses were recently recorded in Kenya. As of 2005 stem rust was still believed to be "largely under control worldwide except in Eastern Africa". But by January 2007 an even more virulent strain had gone across the Red Sea into Yemen. FAO first reported on 5 March 2008 that Ug99 had now spread to major wheat-growing areas in Iran. These countries in North Africa and Middle East consume over 150% of their own wheat production; the failure of this staple crop thus adds a major burden on them. The disease is now expected to spread over China and the Far-East. The strong international collaboration network of research and development that spread disease-resistant strains some 40 years ago and started the Green Revolution, was since slowly starved of research funds because of its own success and is now too atrophied to swiftly react to the new threat.

Soil and Productivity Losses

Sundquist points out that large areas of croplands are lost year after year, due mainly to soil erosion, water depletion and urbanisation. According to him "60,000 km²/year of land becomes so severely degraded that it loses its productive capacity and becomes wasteland", and even more are affected to a lesser extent, adding to the crop supply problem.

Additionally, agricultural production is also lost due to water depletion. Northern China in particular has depleted much of its non-renewables aquifers, which now impacts negatively its crop production.

Urbanisation is another, smaller, difficult to estimate cause of annual cropland reduction.

Rising Levels of Ozone

One possible environmental factor in the food price crisis is rising background levels of ozone in the atmosphere. Plants have been shown to have a high sensitivity to ozone levels, and lower yields of important food crops, such as wheat and soybeans, may have be a result of ozone levels. Ozone levels in the Yangtze Delta were studied for their effect on oilseed rape, a member of the cabbage family that produces one-third of the vegetable oil used in China. Plants grown in chambers that controlled ozone levels exhibited a 10–20 percent reduction in size and weight (biomass) when exposed to elevated ozone. Production of seeds and oil was also reduced.

Rising Prices

From the beginning of 2007 to early 2008, the prices of some of the most basic international food commodities increased dramatically on international markets. The international market price of wheat doubled from February 2007 to February 2008 hitting a record high of over USD$10 a bushel. Rice prices also reached ten year highs. In some nations, milk and meat prices more than doubled, while soy (which hit a 34 year high price in December 2007) and maize prices have increased dramatically.

Total food import bills rose by an estimated 25% for developing countries in 2007. Researchers from the Overseas Development Institute have suggested this problem will be worsened by a likely fall in food aid. As food aid is programmed by budget rather than volume, rising food prices mean that the World Food Programme (WFP) needs an extra $500 million just to sustain the current operations.

To ensure that food remains available for their domestic populations and to combat dramatic price inflation, major rice exporters, such as China, Brazil, India, Indonesia, Vietnam, Cambodia and Egypt, have imposed strict export bans on rice. Conversely, several other nations, including Argentina, Ukraine, Russia, and Serbia have, as well, either imposed high tariffs or blocked the export of wheat and other foodstuffs altogether, driving up prices still further for net food importing nations while trying to isolate their internal markets. Finally,

North Korea, is also suffering from the food crisis (to such extent that a North Korean official was quoted in June '08 with saying "Life is more than difficult. It seems that everyone is going to die").. This nation however is solely relying on food assistance to cope with the crisis. In Developed Countries

United States

The global food price crisis has appeared in the U. S. A. as the rice shortage. The retail prices of food in the U. S. increased four percent according to the Consumer Price Index in 2007, the largest increase in 17 years. The USDA Economic Research Service predicted that prices would increase another three to four percentage points throughout 2008.

In April 2008 Sam's Club instituted a limit on how much long-grain white rice that restaurant and retail customers could purchase due to shortages. Purchases of other types of rice were not restricted. A May 2008 national survey found that food banks and pantries across the U. S. were being forced to cut back on food distribution as 99 percent of respondents reported an increase in the number of people requesting services. Rising food and fuel prices, inadequate food stamp benefits, unemployment, underemployment, and rent or mortgage costs were factors reported as forcing an average of 15-20 percent more people. Compounding this issue, USDA bonus foods have declined by $200 million and local food donations were down nationally about 9 percent over the same period. According to the California Association of Food Banks, which is an umbrella organization of nearly all food banks in the state, food banks are at the "beginning of a crisis."

Impacts on Farmers

If global price movements are transmitted to local markets, farmers in the developing world could benefit from the rising price of food.

According to researchers from the Overseas Development Institute, this may depend on farmers' capacity to respond to changing market conditions. Experience suggests that farmers lack the credit and inputs needed to respond in the short term. In the medium or long term, however, they could benefit, as seen in the Asian green revolution or in many African countries in the recent past.

Unrest and Government Actions in Individual Countries and Regions

The price rises affected parts of Asia and Africa particularly severely with Burkina Faso, Cameroon, Senegal, Mauritania, Cote d'Ivoire, Egypt and Morocco seeing protests and riots in late 2007 and early 2008 over the unavailability of basic food staples. Other countries that have seen food riots or are facing related unrest are: Mexico, Bolivia, Yemen, Uzbekistan, Bangladesh, Pakistan, Sri Lanka, and South Africa.

Bangladesh

10,000 workers rioted close to the Bangladeshi capital Dhaka, smashing cars and buses and vandalising factories in anger at high food prices and low wages. Dozens of people, including at least 20 police officials, were injured in the violence. Ironically, the country achieved food self-sufficiency in 2002, but food prices increased drastically due to the reliance of agriculture on oil and fossil fuels.

Economists estimate 30 million of the country's 150 million people could go hungry.

Brazil

In April 2008, the Brazilian government announced a temporary ban on the export of rice. The ban is intended to protect domestic consumers.

Burkina Faso

One of the earlier food riots took place in Burkina Faso, on February 22, when rioting broke in the country's second and third largest cities over soaring food prices (up to 65 percent increase), sparing however the capital, Ouagadougou, where soldiers were mobilized throughout strategic points. The government promised to lower taxes on food and to release food stocks. Over 100 people were arrested in one of the towns. Related policy actions of Burkinabe government included the removal of customs duty on rice, salt, dairy-based products and baby foods; removal of value added tax on durum wheat, baby foods, soap and edible oils; establishing negotiated with wholesalers prices for sugar, oil and rice; release of food stock; strengthening of community grain banks; food distribution in-kind; reduction of electricity cost, partial payment of utility bills for the poor; enacting special programs for school and hospital feeding, fertilizer distribution and production support. Also, an ban was imposed on exportation of cereals.

Cameroon

Cameroon, the world's fourth largest cocoa producer, saw large scale rioting in late February 2008, in protest against inflating food and fuel prices, as well as the attempt by President Paul Biya to extend his 25-year rule. At least seven people were killed in the worst unrest seen in the country in over fifteen years. This figure was later increased to 24. Part of the government response to the protests was a reduction in import taxes on foods including rice, flour, and fish. The government reached an agreement with retailers by which prices would be lowered in exchange for the reduced import taxes. As of late April 2008, however, reports suggested that prices had not eased and in some cases had even increased.

On April 24, 2008, the government of Cameroon announced a two-year emergency program designed to double Cameroon's food production and achieve food self-sufficiency.

Cote D'Ivoire

On March 31, Cote d'Ivoire's capital Abidjan saw police use tear gas and a dozen protesters injured following food riots that gripped the city. The riots followed dramatic hikes in the price of food and fuel, with the price of beef rising from $1. 68 to $2. 16 per kilogram, and the price of gasoline rising from $1. 44 to $2. 04 per liter, in only three days.

Egypt

In Egypt, a boy was killed from a gunshot to the head after Egyptian police intervened in violent demonstrations over rising food prices that gripped the industrial city of Mahalla on April 8. Large swathes of the population have been hit as food prices, and particularly the price of bread, have doubled over the last several months as a result of producers exploiting a shortage that has existed since 2006.

Ethiopia

Drought and the food price crisis are threatening thousands in Ethiopia.

Haiti

On April 12, 2008, the Haitian Senate voted to dismiss Prime Minister Jacques-Edouard Alexis after violent food riots hit the country.

The food riots caused the death of 5 people. Prices for food items such as rice, beans, fruit and condensed milk have gone up 50 percent in Haiti since late 2007 while the price of fuel has tripled in only two months. Riots broke out in April due to the high prices, and the government had been attempting to restore order by subsidizing a 15 percent reduction in the price of rice. As of February 2010, post-earthquake Port-au-Prince is almost entirely reliant on foreign food aid, some of which ends up in the black markets.

India

Food riots were reported in the Indian state of West Bengal in 2007 over shortages of food. India has banned the export of rice except for Basmati types of rice, which attract a premium price.

Indonesia

Street protests over the price of food took place in Indonesia where food staples and gasoline have nearly doubled in price since January 2008.

Latin America

In April 2008, the Latin American members of the United Nations' Food and Agriculture Organization (FAO) met in Brasilia to confront the issues of high food prices, scarcities and violence that affect the region.

Mexico

The President of Mexico, Felipe Calderon, with industry representatives and members of the Confederation of Industrial Chambers (Concamin), agreed to freeze prices of more than 150 consumer staples, such as coffee, sardines, tuna, oil, soup or tea, among others, until the end of December 2008. The measure was carried out in an attempt to control inflation, which stood at an annual rate of 4. 95%, the highest increase since December 2004.

Mexican baking company Grupo Bimbo, agreed to maintain stable prices of their products, despite of the 20% raise of the production costs.. Daniel Servitje, CEO of this company, announced in the 19th Plenary Meeting of the Mexico–China Business Committee. Bimbo is one of the most important baking companies worldwide, with the recent expansion to China. Bimbo also recently has acquired five bakeries in the United States and Canada.. Servitje also said that their investment plans of this year will be done, but the long term projects, can be changed.

Mozambique

In mid February, rioting that started in the Mozambican rural town of Chokwe and then spread to the capital, Maputo, has resulted in at least four deaths. The riots were reported in the media to have been, at least in part, over food prices and were termed "food riots." A biofuel advocacy publication, however, claimed that these were, in fact, "fuel riots", limited to the rise in the cost of diesel, and argued that the "food riot" characterization worked to fan "anti-biofuels sentiment."

Pakistan

The Pakistan Army has been deployed to avoid the seizure of food from fields and warehouses. This hasn't stopped the food prices from increasing. The new government has been blamed for not managing the countries food stockpiles properly.

Myanmar

Once the world's top rice producer, Myanmar has produced enough rice to feed itself until now. Rice exports dropped over four decades from nearly 4 million tons to only about 40,000 tons last year, mostly due to neglect by Myanmar's ruling generals of infrastructure, including irrigation and warehousing. On 3 May 2008 Cyclone Nargis stripped Myanmar's rice-growing districts, ruining large areas with salt water. U. N. Food and Agriculture Organization estimates that these areas produce 65 percent of the Southeast Asian country's rice.

Worries of long-term food shortages and rationing are rife. The military regime says nothing about the rice crisis, but continues to export rice at the same rate. "... at least the next two harvests are going to be greatly affected and there'll be virtually no output from those areas during that time. So we're likely to see considerable food and rice shortages for the next couple of years. The damage to the economy is going to be profound." said economist and Myanmar expert Sean Turnell, of Australia's Macquarie University.

Panama

In Panama, in response to higher rice prices the government began buying rice at the high market price and selling rice to the public at a lower subsidized price at food kiosks.

Philippines

In the Philippines, the Arroyo government insisted on April 13 that there would be no food riots in the country and that there could be no comparison with Haiti's situation. Chief Presidential Legal Counsel, Sergio Apostol stated that: "Haiti is not trying to solve the problem, while we are doing something to address the issue. We don't have a food shortage. So, no comparison..." Comments by the Justice Secretary, Raul Gonzalez, the following day, that food riots are not far fetched, were quickly rebuked by the rest of the government.

On April 15, the Philippines, the world's largest rice importer, urged China, Japan, and other key Asian nations, to convene an emergency meeting, especially taking issue with those countries' rice export bans. "Free trade should be flowing," Philippine Agriculture

Secretary Arthur Yap stated. In late April 2008, the Philippines government requested that the World Bank exert pressure on rice exporting countries to end export restrictions.

Russia

The Russian government pressured retailers to freeze food prices before key elections for fear of a public backlash against the rising cost of food in October 2007. The freeze ended on May 1, 2008.

Senegal

On 31 March 2008, Senegal had riots in response to the rise in the price of food and fuel. Twenty-four people were arrested and detained in a response that one local human rights group claimed included "torture" and other "unspeakable acts" on the part of the security forces. Further protests took place in Dakar on 26 April 2008.

Somalia

Witnesses and officials in Somalia said thousands of angry Somalis rioted on May 5, 2008 over rising food prices and the collapse of the nation's currency, culminating in clashes with government troops and armed shopkeepers that killed at least five protesters. The protests occurred amid a serious humanitarian emergency due to the Ethiopian war in Somalia.

Tajikistan

The Christian Science Monitor, neweurasia, and other media observers are predicting that a nascent hunger crisis will erupt into a full famine as a consequence of the energy shortages. UN experts announced on 10 October that almost one-third of Tajikistan's 6. 7 million inhabitants may not have enough to eat for the winter of 2008-09.

Yemen

Food riots in southern Yemen that began in late March and continued through early April, saw police stations torched, and roadblocks were set up by armed protesters. The army has deployed tanks and other military vehicles. Although the riots involved thousands of demonstrators over several days and over 100 arrests, officials claimed no fatalities; residents, however, claimed that at least one of the fourteen wounded people has died.

Projections

The UN (FAO) released a study in December 2007 projecting a 49 percent increase in African cereal prices, and 53 percent in European prices, through July 2008. In April 2008, the World Bank, in combination with the International Monetary Fund, announced a series of measures aimed at mitigating the crisis, including increased loans to African farmers and emergency monetary aid to badly affected areas such as Haiti.

According to FAO director Jacques Diouf, however, the World Food Programme needs an immediate cash injection of at least $1700 million, far more than the tens of million-worth in measures already pledged. On 28 April 2008, the United Nations Secretary-General Ban Ki-moon established a Task Force on the Global Food Security Crisis under his chairmanship and composed of the heads of the United Nations specialized agencies, funds and programmes, Bretton Woods institutions and relevant parts of the UN Secretariat to co-ordinate efforts to alleviate the crisis.

Actions by Governments

IFAD is making up to US$200 million available to support poor farmers boost food production in face of the global food crisis.

On May 2, 2008 U. S. President George W. Bush said he was asking Congress to approve an extra $770 million funding for international food aid. On October 16, 2008, former US president Bill Clinton scolded the bipartisan coalition in Congress that killed the idea of making some aid donations in cash rather than in food.

The release of Japan's rice reserves onto the market may bring the rice price down significantly. As of May 16, anticipation of the move had already lowered prices by 14% in a single week.

On April 30, 2008 Thailand announced the creation of the Organization of Rice Exporting Countries (OREC) with the potential to develop a price-fixing cartel for rice. This is seen by some as an action to capitalise on the crisis.

In June 2008 the Food and Agriculture Organization hosted a High-Level Conference on World Food Security, in which $1. 2 billion in food aid was committed for the 75 million people in 60 countries hardest hit by rising food prices.

In June 2008, a sustained commitment from the G8 was called for by some humanitarian organizations.

On October 23, 2008, Associated Press reported the following:

> *"Former President Clinton told a U. N. gathering Thursday [Oct 16, 2008] that the global food crisis shows "we all blew it, including me," by treating food crops "like*

colour TVs" instead of as a vital commodity for the world's poor.... Clinton criticized decades of policymaking by the World Bank, the International Monetary Fund and others, encouraged by the U. S., that pressured Africans in particular into dropping government subsidies for fertilizer, improved seed and other farm inputs as a requirement to get aid. Africa's food self-sufficiency declined and food imports rose. Now skyrocketing prices in the international grain trade — on average more than doubling between 2006 and early 2008 — have pushed many in poor countries deeper into poverty."

We need the World Bank, the IMF, all the big foundations, and all the governments to admit that, for 30 years, we all blew it, including me when I was President. We were wrong to believe that food was like some other product in international trade, and we all have to go back to a more responsible and sustainable form of agriculture.

– Former US President Bill Clinton, Speech at United Nations World Food Day, October 16, 2008

Food Price Decreases

In December 2008, the global economic slowdown, decreasing oil prices, and speculation of decreased demand for commodities worldwide brought about sharp decreases in the price of staple crops from their earlier highs. Corn prices on the Chicago Board of Trade dropped from US $7. 99 per bushel in June to US $3. 74 per bushel in mid-December; wheat and rice prices experienced similar decreases. The UN's Food and Agriculture Organization, however, warned against "a false sense of security", noting that the credit crisis could cause farmers to reduce plantings.. FAO has convened a World Summit on Food Security at its headquarters in Rome in November, noting that food prices remain high in developing countries and that the global food security situation has worsened.

Late 2000s Recession

The late-2000s recession (or the Great Recession) is an economic recession that began in the United States in December 2007 (and with much greater intensity since September 2008, according to the National Bureau of Economic Research). It spread to much of the industrialized world, and has caused a pronounced deceleration of economic activity. This global recession has been taking place in an economic environment characterized by various imbalances and was sparked by the outbreak of the financial crisis of 2007–2010. Although the late-2000s recession has at times been referred to as "the Great Recession," this same phrase has been used to refer to every recession of the several preceding decades. In July 2009, it was announced that a growing number of economists believed that the recession may have ended.

The financial crisis has been linked to reckless and unsustainable lending practices compounded by government intervention and the growing trend of securitization of real estate mortgages in the United States. The US mortgage-backed securities, which had risks that were hard to assess, were marketed around the world. A more broad based credit boom fed a global speculative bubble in real estate and equities, which served to reinforce the risky lending practices. The precarious financial situation was made more difficult by a sharp increase in oil and food prices.

The emergence of Sub-prime loan losses in 2007 began the crisis and exposed other risky loans and over-inflated asset prices. With loan losses mounting and the fall of Lehman Brothers on September 15, 2008, a major panic broke out on the inter-bank loan market. As share and housing prices declined, many large and well established investment and commercial banks in the United States and Europe suffered huge losses and even faced bankruptcy, resulting in massive public financial assistance.

A global recession has resulted in a sharp drop in international trade, rising unemployment and slumping commodity prices. In December 2008, the National Bureau of Economic Research (NBER) declared that the United States had been in recession since December 2007. Several economists have predicted that recovery may not appear until 2011 and that the recession will be the worst since the Great Depression of the 1930s. The conditions leading up to the crisis, characterised by an exorbitant rise in asset prices and associated boom in economic demand, are considered a result of the extended period of easily available credit, inadequate regulation and oversight, or increasing inequality.

The recession has renewed interest in Keynesian economic ideas on how to combat recessionary conditions. Fiscal and monetary policies have been significantly eased to stem the recession and financial risks. Economists advise that the stimulus should be withdrawn as soon as the economies recover enough to "chart a path to sustainable growth".

Unfortunately, Keynesian stimulus efforts, while providing a short-term bump, also increase the public debt, which can lead to funding problems, as became clear in 2010 with the European sovereign debt crisis threatening to cause a “double-dip recession”.

Pre-recession Economic Imbalances

The onset of the economic crisis took most people by surprise. A 2009 paper identifies twelve economists and commentators who, between 2000 and 2006, predicted a recession based on the collapse of the then booming housing market in the U. S: Dean Baker, Wynne Godley, Fred Harrison, Michael Hudson, Eric Janszen, Steve Keen, Jakob Brochner Madsen & Jens Kjaer Sorensen, Kurt Richebacher, Nouriel Roubini, Peter Schiff and Robert Shiller. Among the various imbalances in which the US monetary policy contributed by excessive money creation, leading to negative household savings and a huge US trade deficit, dollar volatility and public deficits, a focus can be made on the following ones:

Commodity Boom

The decade of the 2000s saw a global explosion in prices, focused especially in commodities and housing, marking an end to the commodities recession of 1980–2000. In 2008, the prices of many commodities, notably oil and food, rose so high as to cause genuine economic damage, threatening stagflation and a reversal of globalization.

In January 2008, oil prices surpassed $100 a barrel for the first time, the first of many price milestones to be passed in the course of the year. In July 2008, oil peaked at $147. 30 a barrel and a gallon of gasoline was more than $4 across most of the U. S. A. The economic contraction in the fourth quarter of 2008 caused a dramatic drop in demand and prices fell below $35 a barrel at the end of the year. Some believe that this oil price spike was the product of Peak Oil. There is concern that if the economy was to improve, oil prices might return to pre-recession levels.

The food and fuel crises were both discussed at the 34th G8 summit in July 2008.

Sulfuric acid (an important chemical commodity used in processes such as steel processing, copper production and bioethanol production) increased in price 3. 5-fold in less than 1 year while producers of sodium hydroxide have declared force majeure due to flooding, precipitating similarly steep price increases.

In the second half of 2008, the prices of most commodities fell dramatically on expectations of diminished demand in a world recession.

Housing Bubble

By 2007, real estate bubbles were still under way in many parts of the world, especially in the United States, United Kingdom, United Arab Emirates, Italy, Australia, New Zealand, Ireland, Spain, France, Poland, South Africa, Israel, Greece, Bulgaria, Croatia, Canada, Norway, Singapore, South Korea, Sweden, Finland, Argentina, Baltic states, India, Romania, Russia, Ukraine and China. U. S. Federal Reserve Chairman Alan Greenspan said in mid-2005 that "at a minimum, there's a little 'froth' (in the U. S. housing market)... it's hard not to see that there are a lot of local bubbles". The *Economist* magazine, writing at the same time, went further, saying "the worldwide rise in house prices is the biggest bubble in history". Real estate bubbles are (by definition of the word "bubble") followed by a price decrease (also known as a housing price crash) that can result in many owners holding negative equity (a mortgage debt higher than the current value of the property).

Inflation

In February 2008, Reuters reported that global inflation was at historic levels, and that domestic inflation was at 10–20 year highs for many nations. "Excess money supply around the globe, monetary easing by the Fed to tame financial crisis, growth surge supported by easy monetary policy in Asia, speculation in commodities, agricultural failure, rising cost of imports from China and rising demand of food and commodities in the fast growing emerging markets," have been named as possible reasons for the inflation.

In mid-2007, IMF data indicated that inflation was highest in the oil-exporting countries, largely due to the unsterilized growth of foreign exchange reserves, the term "unsterilized" referring to a lack of monetary policy operations that could offset such a foreign exchange intervention in order to maintain a country's monetary policy target. However, inflation was also growing in countries classified by the IMF as "non-oil-exporting LDCs" (Least Developed Countries) and "Developing Asia", on account of the rise in oil and food prices.

Inflation was also increasing in the developed countries, but remained low compared to the developing world.

Causes

1. Central banks' gold reserves – $0. 845 tn.
2. M0 (paper money) —-$3. 9 tn.

3. traditional (fractional reserve) banking assets – $39 tn.
4. shadow banking assets – $62 tn.
5. other assets – $290 tn.
6. Bail-out money (early 2009) – $1. 9 tn.

Debate over Origins

The central debate about the origin has been focused on the respective parts played by the public monetary policy (in the US notably) and by private financial institutions practices.

On October 15, 2008, Anthony Faiola, Ellen Nakashima, and Jill Drew wrote a lengthy article in *The Washington Post* titled, "What Went Wrong". In their investigation, the authors claim that former Federal Reserve Board Chairman Alan Greenspan, Treasury Secretary Robert Rubin, and SEC Chairman Arthur Levitt vehemently opposed any regulation of financial instruments known as derivatives.

They further claim that Greenspan actively sought to undermine the office of the Commodity Futures Trading Commission, specifically under the leadership of Brooksley E. Born, when the Commission sought to initiate regulation of derivatives. Ultimately, it was the collapse of a specific kind of derivative, the mortgage-backed security, that triggered the economic crisis of 2008.

While Greenspan's role as Chairman of the Federal Reserve has been widely discussed (the main point of controversy remains the lowering of Federal funds rate at only 1% for more than a year which, according to the Austrian School of economics, allowed huge amounts of "easy" credit-based money to be injected into the financial system and thus create an unsustainable economic boom), there is also the argument that Greenspan actions in the years 2002–2004 were actually motivated by the need to take the U. S. economy out of the early 2000s recession caused by the bursting of the dot-com bubble — although by doing so he did not help avert the crisis, but only postpone it.

Some economists-those of the Austrian school and those predicting the recession such as Steve Keen-claim that the ultimate point of origin of the great financial crisis of 2007–2009 can be traced back to an extremely indebted US economy. The collapse of the real estate market in 2006 was the close point of origin of the crisis.

The failure rates of subprime mortgages were the first symptom of a credit boom tuned to bust and of a real estate shock. But large default rates on subprime mortgages cannot account for the severity of the crisis. Rather, low-quality mortgages acted as an accelerant to the fire that spread through the entire financial system. The latter had become fragile as a result of several factors that are unique to this crisis: the transfer of assets from the balance sheets of banks to the markets, the creation of complex and opaque assets, the failure of ratings agencies to properly assess the risk of such assets, and the application of fair value accounting. To these novel factors, one must add the now standard failure of regulators and supervisors in spotting and correcting the emerging weaknesses.

Subprime Lending as a Cause

Based on the assumption that subprime lending precipitated the crisis, some have argued that the Clinton Administration may be partially to blame, while others have pointed to the passage of the Gramm-Leach-Bliley Act by the 106th Congress, and over-leveraging by banks and investors eager to achieve high returns on capital.

Others take full credit for de-regulating the Banking Industry. In November 1999, Phil Gramm, Republican Senator from Texas, took full credit for the Gramm-Leach-Bliley Act with a Press Release from the Senate Banking and Finance Committee: "I am proud to be here because this is an important bill; it is a deregulatory bill. I believe that is the wave of the future, and I am awfully proud to have been a part of making it a reality."

Some believe the roots of the crisis can be traced directly to subprime lending by Fannie Mae and Freddie Mac, which are government sponsored entities. The New York Times published an article that reported the Clinton Administration pushed for subprime lending: "Fannie Mae, the nation's biggest underwriter of home mortgages, has been under increasing pressure from the Clinton Administration to expand mortgage loans among low and moderate income people" (NYT, 30 September 1999).

In 1995, the administration also tinkered with Carter's Community Reinvestment Act of 1977 by regulating and strengthening the anti-redlining procedures. It is felt by many that this was done to help boost a stagnated home ownership figure that had hovered around 65% for many years.

The result was a push by the administration for greater investment, by financial institutions, into riskier loans. In a 2000 United States Department of the Treasury study of lending trends for 305 cities from 1993 to 1998 it was shown that $467 billion of mortgage credit poured out of CRA-covered lenders into low-and mid-level income borrowers and neighbourhoods.

Government Deregulation as a Cause

In 1992, the 102nd Congress under the George H. W. Bush administration weakened regulation of Fannie Mae and Freddie Mac with the goal of making available more money for the issuance of home loans. The Washington Post wrote: "Congress also wanted to free up money for Fannie Mae and Freddie Mac to buy mortgage loans and specified that the pair would be required to keep a much smaller share of their funds on hand than other financial institutions. Whereas banks that held $100 could spend $90 buying mortgage loans, Fannie Mae and Freddie Mac could spend $97. 50 buying loans. Finally, Congress ordered that the companies be required to keep more capital as a cushion against losses if they invested in riskier securities. But the rule was never set during the Clinton administration, which came to office that winter, and was only put in place nine years later."

Others have pointed to deregulation efforts as contributing to the collapse. In 1999, the 106th Congress, under Bill Clinton, passed the Gramm-Leach-Bliley Act, which repealed part of the Glass-Steagall Act of 1933. This repeal has been criticized by some for having contributed to the proliferation of the complex and opaque financial instruments which are at the heart of the crisis. However, some economists object to singling out the repeal of Glass-Steagall for criticism. Brad DeLong, a former advisor to President Clinton and economist at the University of California, Berkeley and Tyler Cowen of George Mason University have both argued that the Gramm-Leach-Bliley Act softened the impact of the crisis by allowing for mergers and acquisitions of collapsing banks as the crisis unfolded in late 2008.

Over-Leveraging, Credit Default Swaps and Collateralized Debt Obligations as Causes

Another probable cause of the crisis—and a factor that unquestionably amplified its magnitude—was widespread miscalculation by banks and investors of the level of risk inherent in the unregulated Collateralized debt obligation and Credit Default Swap markets. Under this theory, banks and investors systematized the risk by taking advantage of low interest rates to borrow tremendous sums of money that they could only pay back if the housing market continued to increase in value.

According to an article published in *Wired*, the risk was further systematized by the use of David X. Li's Gaussian copula model function to rapidly price Collateralized debt obligations based on the price of related Credit Default Swaps. Because it was highly tractable, it rapidly came to be used by a huge percentage of CDO and CDS investors, issuers, and rating agencies. According to one wired com article: "Then the model fell apart. Cracks started appearing early on, when financial markets began behaving in ways that users of Li's formula hadn't expected. The cracks became full-fledged canyons in 2008—when ruptures in the financial system's foundation swallowed up trillions of dollars and put the survival of the global banking system in serious peril... Li's Gaussian copula formula will go down in history as instrumental in causing the unfathomable losses that brought the world financial system to its knees."

The pricing model for CDOs clearly did not reflect the level of risk they introduced into the system. It has been estimated that the "from late 2005 to the middle of 2007, around $450bn of CDO of ABS were issued, of which about one third were created from risky mortgage-backed bonds... [o]ut of that pile, around $305bn of the CDOs are now in a formal state of default, with the CDOs underwritten by Merrill Lynch accounting for the biggest pile of defaulted assets, followed by UBS and Citi." The average recovery rate for high quality CDOs has been approximately 32 cents on the dollar, while the recovery rate for mezzanine CDO's has been approximately five cents for every dollar. These massive, practically unthinkable, losses have dramatically impacted the balance sheets of banks across the globe, leaving them with very little capital to continue operations.

Credit Creation as a Cause

The Austrian School of Economics proposes that the crisis is an excellent example of the Austrian Business Cycle Theory, in which credit created through the policies of central banking gives rise to an artificial boom, which is inevitably followed by a bust. This perspective argues that the monetary policy of central banks creates excessive quantities of cheap credit by setting interest rates below where they would be set by a free market. This easy availability of credit inspires a bundle of malinvestments, particularly on long term projects such as housing and capital assets, and also spurs a consumption boom as incentives to save are diminished. Thus an unsustainable boom arises, characterized by malinvestments and over consumption.

But the created credit is not backed by any real savings nor is in response to any change in the real economy, hence, there are physically not enough resources to finance either the malinvestments or the consumption rate indefinitely. The bust occurs when investors collectively realize their mistake. This happens usually some time after interest rates rise again. The liquidation of the malinvestments and the consequent reduction in consumption throw the economy into a recession, whose severity mirrors the scale of the boom's excesses.

The Austrian School argues that the conditions previous to the crisis of the late 2000s correspond exactly to the scenario described above. The central bank of the United States, led by Federal Reserve Chairman Alan Greenspan, kept interest rates very low for a long period of time to blunt the recession of the early 2000s. The resulting mal investment and over consumption of investors and consumers prompted the development of a housing bubble that ultimately burst, precipitating the financial crisis. This crisis, together with sudden and necessary diverging and cutbacks by consumers, businesses and banks, led to the recession. Austrian Economists argue further that while they probably affected the nature and severity of the crisis, factors such as a lack of regulation, the Community Reinvestment Act, and entities such as Fannie Mae and Freddie Mac are insufficient by themselves to explain it.

Austrian economists argue that the history of the yield curve from 2000 through 2007 illustrates the role that credit creation by the Federal Reserve may have played in the on-set of the financial crisis in 2007 and 2008. The yield curve (also known as the term structure of interest rates) is the shape formed by a graph showing US Treasury Bill or Bond interest rates on the vertical axis and time to maturity on the horizontal axis. When short-term interest rates are lower than long-term interest rates the yield curve is said to be "positively sloped". When short-term interest rates are higher than long-term interest rates the yield curve is said to be "inverted". When long term and short term interest rates are equal the yield curve is said to be "flat". The yield curve is believed by some to be a strong predictor of recession (when inverted) and inflation (when positively sloped). However, the yield curve is believed to act on the real economy with a lag of 1 to 3 years.

A positively sloped yield curve allows Primary Dealers (such as large investment banks) in the Federal Reserve system to fund themselves with cheap short term money while lending out at higher long-term rates. This strategy is profitable so long as the yield curve remains positively sloped. However, it creates a liquidity risk if the yield curve were to become inverted and banks would have to refund themselves at expensive short term rates while losing money on longer term loans.

The narrowing of the yield curve from 2004 and the inversion of the yield curve during 2007 resulted (with the expected 1 to 3 year delay) in a bursting of the housing bubble and a wild gyration of commodities prices as moneys flowed out of assets like housing or stocks and sought safe haven in commodities. The price of oil rose to over $140 dollars per barrel in 2008 before plunging as the financial crisis began to take hold in late 2008.

Other observers have doubted the role that the yield curve plays in controlling the business cycle. In a May 24, 2006 story CNN Money reported:

> *"... in recent comments, Fed Chairman Ben Bernanke repeated the view expressed by his predecessor Alan Greenspan that an inverted yield curve is no longer a good indicator of a recession ahead."*

Oil Prices

Economist James D. Hamilton has argued that the increase in oil prices in the period of 2007 through 2008 was a significant cause of the recession. He evaluated several different approaches to estimating the impact of oil price shocks on the economy, including some methods that had previously shown a decline in the relationship between oil price shocks and the overall economy. All of these methods "support a common conclusion; had there been no increase in oil prices between 2007:Q3 and 2008:Q2, the US economy would not have been in a recession over the period 2007:Q4 through 2008:Q3." Hamilton's own model, a time-series econometric forecast based on data up to 2003, showed that the decline in GDP could have been successfully predicted to almost its full extent given knowledge of the price of oil. The results imply that oil prices were entirely responsible for the recession; however, Hamilton himself acknowledged that this was probably not the case but maintained that it showed that oil price increases made a significant contribution to the downturn in economic growth.

Other Claimed Causes

Many libertarians, including Congressman and former 2008 Presidential candidate Ron Paul and Peter Schiff in his book *Crash Proof*, claim to have predicted the crisis prior to its occurrence. Schiff also made a speech in 2006 in which he predicted the failure of Fannie and Freddie. They are critical of theories that the free market caused the crisis and instead argue that the Federal Reserve's expansionary monetary policy and the Community Reinvestment Act are the primary causes of the crisis. Alan Greenspan, former Federal Reserve chairman, has said he was partially wrong to oppose regulation of the markets, and expressed "shocked disbelief" at the failure of self interest, alone, to manage risk in the markets.

An empirical study by John B. Taylor concluded that the crisis was:

(1) caused by excess monetary expansion;

(2) prolonged by an inability to evaluate counter-party risk due to opaque financial statements; and

(3) worsened by the unpredictable nature of government's response to the crisis.

It has also been debated that the root cause of the crisis is overproduction of goods caused by globalization (and especially vast investments in countries such as China and India by western multinational companies over the past 15–20 years, which greatly increased global industrial output at a reduced cost). Overproduction tends to cause deflation and signs of deflation were evident in October and November 2008, as commodity prices tumbled and the Federal Reserve was lowering its target rate to an all-time-low 0. 25%. On the other hand, Professor Herman Daly suggests that it is not actually an economic crisis, but rather a crisis of overgrowth beyond sustainable ecological limits. This reflects a claim made in the 1972 book *Limits to Growth*, which stated that without major deviation from the policies followed in the 20th century, a permanent end of economic growth could be reached sometime in the first two decades of the 21st century, due to gradual depletion of natural resources.

In laissez-faire capitalism, financial institutions would be risk-averse because failure would result in liquidation. But the Federal Reserve's 1984 rescue of Continental Illinois and the 1998 rescue of the Long-Term Capital Management hedge fund, among others, showed that institutions which failed to exercise due diligence could reasonably expect to be protected from the consequences of their mistakes. The belief that they could not be allowed to fail created a moral hazard which was a contributing factor to the late-2000s recession.

Effects

Overview

The late-2000s recession is shaping up to be the worst post-World War II contraction on record:

- Real gross domestic product (GDP) began contracting in the third quarter of 2008, and by early 2009 was falling at an annualized pace not seen since the 1950s.
- Capital investment, which was in decline year-on-year since the final quarter of 2006, matched the 1957–58 post war record in the first quarter of 2009. The pace of collapse in residential

investment picked up speed in the first quarter of 2009, dropping 23. 2% year-on-year, nearly four percentage points faster than in the previous quarter.

- Domestic demand, in decline for five straight quarters, is still three months shy of the 1974–75 record, but the pace – down 2. 6% per quarter vs. 1. 9% in the earlier period – is a record-breaker already.

Trade and Industrial Production

In middle-October 2008, the Baltic Dry Index, a measure of shipping volume, fell by 50% in one week, as the credit crunch made it difficult for exporters to obtain letters of credit.

In February 2009, *The Economist* claimed that the financial crisis had produced a "manufacturing crisis", with the strongest declines in industrial production occurring in export-based economies.

In March 2009, Britain's *Daily Telegraph* reported the following declines in industrial output, from January 2008 to January 2009: Japan "31%, Korea "26%, Russia "16%, Brazil "15%, Italy "14%, Germany "12%. Some analysts even say the world is going through a period of deglobalization and protectionism after years of increasing economic integration.

Sovereign funds and private buyers from the Middle East and Asia, including China, are increasingly buying in on stakes of European and U. S. businesses, including industrial enterprises. Due to the global recession they are available at a low price. The Chinese government has concentrated on natural-resource deals across the world, securing supplies of oil and minerals.

Pollution

According to the International Energy Agency man-made greenhouse gas emissions will decrease by 3% in 2009, mainly as a result of the financial crisis. Previously emissions had been rising by around 3% per year. The drop in emissions is only the 4th to occur in 50 years.

Unemployment

The International Labour Organization (ILO) predicted that at least 20 million jobs will have been lost by the end of 2009 due to the crisis — mostly in "construction, real estate, financial services, and the auto sector" — bringing world unemployment above 200 million for the first time. The number of unemployed people worldwide could increase by more than 50 million in 2009 as the global recession intensifies, the ILO has forecast.

In December 2007, the U. S. unemployment rate was 4. 9%. By October 2009, the unemployment rate had risen to 10. 1%. A broader measure of unemployment (taking into account marginally attached workers, those employed part time for economic reasons, and discouraged workers) was 16. 3%. In July 2009, fewer jobs were lost than expected, dipping the unemployment rate from 9. 5% to 9. 4%. Even fewer jobs were lost in August, 216,000, recorded as the lowest number of jobs since September 2008, but the unemployment rate rose to 9. 7%. In October 2009, news reports announced that some employers who cut jobs due to the recession are beginning to hire them back. More recently, economists announced in January 2010 that economic growth in the U. S. resumed in the fourth quarter of 2009, and some have predicted that limited job growth will begin in the spring of 2010.

The average numbers for European Union nations are similar to the US ones. Some European countries have been hit by recession very hard, for instance Spain's unemployment rate reached 18. 7% (37% for youths) in May 2009 — the highest in the eurozone.

The rise of advanced economies in Brazil, India, and China increased the total global labour pool dramatically. Recent improvements in communication and education in these countries has allowed workers in these countries to compete more closely with workers in traditionally strong economies, such as the United States. This huge surge in labour supply has provided downward pressure on wages and contributed to unemployment.

Financial Markets

For a time, major economies of the 21st century were believed to have begun a period of decreased volatility, which was sometimes dubbed The Great Moderation, because many economic variables appeared to have achieved relative stability. The return of commodity, stock market, and currency value volatility are regarded as indications that the concepts behind the Great Moderation were guided by false beliefs.

January 2008 was an especially volatile month in world stock markets, with a surge in implied volatility measurements of the US-based S&P 500 index, and a sharp decrease in non-U. S. stock market prices on Monday, January 21, 2008 (continuing to a lesser extent in some markets on January 22). Some headline writers and a general news columnist called January 21 "Black Monday" and referred to a "global shares crash," though the effects were quite different in different markets. The effects of these events were also felt on the Shanghai

Composite Index in China which lost 5. 14 percent, most of this on financial stocks such as Ping An Insurance and China Life which lost 10 and 8. 76 percent respectively. Investors worried about the effect of a recession in the US economy would have on the Chinese economy. Citigroup estimates due to the number of exports from China to America a one percent drop in US economic growth would lead to a 1. 3 percent drop in China's growth rate.

There were several large Monday declines in stock markets world wide during 2008, including one in January, one in August, one in September, and another in early October. As of October 2008, stocks in North America, Europe, and the Asia-Pacific region had all fallen by about 30% since the beginning of the year. The Dow Jones Industrial Average had fallen about 37% since January 2008.

The simultaneous multiple crises affecting the US financial system in mid-September 2008 caused large falls in markets both in the US and elsewhere. Numerous indicators of risk and of investor fear (the TED spread, Treasury yields, the dollar value of gold) set records.

Russian markets, already falling due to declining oil prices and political tensions with the West, fell over 10% in one day, leading to a suspension of trading, while other emerging markets also exhibited losses.

On September 18, UK regulators announced a temporary ban on short-selling of financial stocks. On September 19 the U. S. Securities and Exchange Commission (SEC) followed by placing a temporary ban of short-selling stocks of 799 specific financial institutions. In addition, the SEC made it easier for institutions to buy back shares of their institutions. The action is based on the view that short selling in a crisis market undermines confidence in financial institutions and erodes their stability.

On September 22, the Australian Securities Exchange (ASX) delayed opening by an hour after a decision was made by the Australian Securities and Investments Commission (ASIC) to ban all short selling on the ASX. This was revised slightly a few days later.

As is often the case in times of financial turmoil and loss of confidence, investors turned to assets which they perceived as tangible or sustainable. The price of gold rose by 30% from middle of 2007 to end of 2008. A further shift in investors' preference towards assets like precious metals or land is discussed in the media.

In March 2009, Blackstone Group CEO Stephen Schwarzman said that up to 45% of global wealth had been destroyed in little less than a year and a half.

Travel

According to Zagat's 2009 U. S. Hotels, Resorts & Spas survey, business travel has decreased in the past year as a result of the recession. 30% of travellers surveyed stated they travel less for business today while only 21% of travellers stated that they travel more. Reasons for the decline in business travel include company travel policy changes, personal economics, economic uncertainty and high airline prices. Hotels are responding to the downturn by dropping rates, ramping up promotions and negotiating deals for both business travellers and tourists.

According to the World Tourism Organization, international travel suffered a strong slowdown beginning in June 2008, and this declining trend intensified during 2009 resulting in a reduction from 922 million international tourist arrivals in 2008 to 880 million visitors in 2009, representing a worldwide decline of 4%, and an estimated 6% decline in international tourism receipts. The decline caused by the recession was further exacerbated in some countries due to the outbreak of the AH1N1 virus.

Insurance

A February 2009 study on the main British insurers showed that most of them do not plan to raise their insurance premiums for the year 2009, in spite of the prediction of a 20% raise made by *The Daily Telegraph* and *The Daily Mirror*. However, it is expected that the capital liquidity will become an issue and determine increases, having their capital tied up in investments yielding smaller dividends, corroborated with the £644 million underwriting losses suffered in 2007.

Small-business Lending

New York Times reported that the U. S. Treasury Department found a sizeable decrease in small-business lending by the 22 largest bank recipients of federal bailout money. The banks reduced their small-business lending by US$12. 5 billion, a decline of 4. 6 percent during a seven-month period ended in November 2009. During that time, the two biggest small-business lenders, Wells Fargo and Bank of America reduced their lending to small-business by 4. 4 percent and 6. 2 percent, respectively. Bank of America explained that about half of the decline was attributable to decrease demand, and a decline in sales and creditworthiness among small businesses furthered the drop.

Countries Most Affected

The crisis affected all countries in some ways, but certain countries were vastly affected more than others. By measuring currency devaluation, equity market decline, and the rise in sovereign bond spreads, a picture of financial devastation emerges. Since these three indicators show financial weakness, taken together, they capture the impact of the crisis. The Carnegie Endowment for International Peace reports in its International Economics Bulletin that Ukraine, as well as Argentina and Jamaica, are the countries most deeply affected by the crisis. Other severely affected countries are Ireland, Russia, Mexico, Hungary, the Baltic states, United States and United Kingdom. By contrast, China, Japan, India, Iran, Peru and Australia are "among the least affected."

Political Instability Related to the Economic Crisis

In December 2008, Greece experienced extensive civil unrest that continued into January and then again in late February many Greeks took part in a massive general strike because of the economic situation and shut down schools, airports, and many other services in Greece. In January 2009, the government leaders of Iceland were forced to call elections two years early after the people of Iceland staged mass protests and clashed with the police due to the government's handling of the economy. Hundreds of thousands protested in France against President Sarkozy's economic policies, but it is important to remember that such protests, although rare in countries such as the USA, are quite a common phenomenon in Europe, even in periods of growth (major strikes in France in 1968, and many others occurring almost every five to ten years). Prompted by the financial crisis in Latvia, the opposition and trade unions there organized a rally against the cabinet of premier Ivars Godmanis. The rally gathered some 10–20 thousand people. In the evening, the rally turned into a riot. The crowd moved to the building of the parliament and attempted to force their way into it, but were repelled by the state's police. Police and protesters also clashed in Lithuania. In addition to various levels of unrest in Europe, Asian countries have also seen various degrees of protest. Communists and others rallied in Moscow to protest the Russian government's economic plans. Protests have also occurred in China as demands from the West for exports were dramatically reduced and unemployment increased.

Beginning February 26, 2009, an Economic Intelligence Briefing was added to the daily intelligence briefings prepared for the President of the United States. This addition reflected the assessment of United States intelligence agencies that the global financial crisis presented a serious threat to international stability. In March 2009, British think tank Economist Intelligence Unit published a special report titled 'Manning the barricades' in which it estimated "who's at risk as deepening economic distress foments social unrest". The Report envisioned the next two years filled with great social upheavals, disrupted economies and toppled governments around the globe.

Business Week in March 2009 stated that global political instability is rising fast due to the global financial crisis and is creating new challenges that need managing. The Associated Press reported in March 2009 that: United States "Director of National Intelligence Dennis Blair has said the economic weakness could lead to political instability in many developing nations." Even some developed countries are seeing political instability.

NPR reports that David Gordon, a former intelligence officer who now leads research at the Eurasia Group, said: "Many, if not most, of the big countries out there have room to accommodate economic downturns without having large-scale political instability if we're in a recession of normal length. If you're in a much longer-run downturn, then all bets are off."

> *"The recent wave of popular unrest was not confined to Eastern Europe. Ireland, Iceland, France, the U. K. and Greece also experienced street protests, but many Eastern European governments seem more vulnerable as they have limited policy options to address the crisis and little or no room for fiscal stimulus due to budgetary or financing constrains. Deeply unpopular austerity measures, including slashed public wages, tax hikes and curbs on social spending will keep fanning public discontent in the Baltic states, Hungary and Romania. Dissatisfaction linked to the economic woes will be amplified in the countries where governments have been weakened by high-profile corruption and fraud scandals (Latvia, Lithuania, Hungary, Romania, the UK and Bulgaria)."*

Policy Responses

The financial phase of the crisis led to emergency interventions in many national financial systems. As the crisis developed into genuine recession in many major economies, economic stimulus meant to revive economic growth became the most common policy tool. After having implemented rescue plans for the banking system, major developed and emerging countries announced plans to relieve their economies. In particular, economic stimulus plans were announced in China, the United States, and the European Union. Bailouts of failing or threatened businesses were carried out or discussed in the USA, the EU, and India. In the final quarter of 2008, the financial crisis saw the G-20 group of major economies assume a new significance as a focus of economic and financial crisis management.

United States Policy Responses

The Federal Reserve, Treasury, and Securities and Exchange Commission took several steps on September 19 to intervene in the crisis. To stop the potential run on money market mutual funds, the Treasury also announced on September 19 a new $50 billion program to insure the investments, similar to the Federal Deposit Insurance Corporation (FDIC) program. Part of the announcements included temporary exceptions to section 23A and 23B (Regulation W), allowing financial groups to more easily share funds within their group. The exceptions would expire on January 30, 2009, unless extended by the Federal Reserve Board. The Securities and Exchange Commission announced termination of short-selling of 799 financial stocks, as well as action against naked short selling, as part of its reaction to the mortgage crisis.

Market Volatility Within US 401(k) and Retirement Plans

The US Pension Protection Act of 2006 included a provision which changed the definition of Qualified Default Investments (QDI) for retirement plans from stable value investments, money market funds, and cash investments to investments which expose an individual to appropriate levels of stock and bond risk based on the years left to retirement. The Act required that Plan Sponsors move the assets of individuals who had never actively elected their investments and had their contributions in the default investment option. This meant that individuals who had defaulted into a cash fund with little fluctuation or growth would soon have their account balances moved to much more aggressive investments.

Starting in early 2008, most US employer-sponsored plans sent notices to their employees informing them that the plan default investment was changing from a cash/stable option to something new, such as a retirement date fund which had significant market exposure. Most participants ignored these notices until September and October, when the market crash was on every news station and media outlet. It was then that participants called their 401(k) and retirement plan providers and discovered losses in excess of 30% in some cases. Call centres for 401(k) providers experienced record call volume and wait times, as millions of inexperienced investors struggled to understand how their investments had been changed so fundamentally without their explicit consent, and reacted in a panic by liquidating everything with any stock or bond exposure, locking in huge losses in their accounts.

Due to the speculation and uncertainty in the market, discussion forums filled with questions about whether or not to liquidate assets and financial gurus were swamped with questions about the right steps to take to protect what remained of their retirement accounts. During the third quarter of 2008, over $72 billion left mutual fund investments that invested in stocks or bonds and rushed into Stable Value investments in the month of October. Against the advice of financial experts, and ignoring historical data illustrating that long-term balanced investing has produced positive returns in all types of markets, investors with decades to retirement instead sold their holdings during one of the largest drops in stock market history.

Loans to Banks for Asset-backed Commercial Paper

During the week ending September 19, 2008, money market mutual funds had begun to experience significant withdrawals of funds by investors. This created a significant risk because money market funds are integral to the ongoing financing of corporations of all types. Individual investors lend money to money market funds, which then provide the funds to corporations in exchange for corporate short-term securities called asset-backed commercial paper (ABCP). However, a potential bank run had begun on certain money market funds. If this situation had worsened, the ability of major corporations to secure needed short-term financing through ABCP issuance would have been significantly affected. To assist with liquidity throughout the system, the US Treasury and Federal Reserve Bank announced that banks could obtain funds via the Federal Reserve's Discount Window using ABCP as collateral.

Legislation

The Secretary of the United States Treasury, Henry Paulson and President George W. Bush proposed legislation for the government to purchase up to US$700 billion of "troubled mortgage-related assets" from financial firms in hopes of improving confidence in the mortgage-backed securities markets and the financial firms participating in it.

Discussion, hearings and meetings among legislative leaders and the administration later made clear that the proposal would undergo significant change before it could be approved by Congress. On October 1, a revised compromise version was approved by the Senate with a 74–25 vote. The bill, HR1424 was passed by the House on October 3,

2008 and signed into law. The first half of the bailout money was primarily used to buy preferred stock in banks instead of troubled mortgage assets.

In January 2009, the Obama administration announced a stimulus plan to revive the economy with the intention to create or save more than 3. 6 million jobs in two years. The cost of this initial recovery plan was estimated at 825 billion dollars (5. 8% of GDP). The plan included 365. 5 billion dollars to be spent on major policy and reform of the health system, 275 billion (through tax rebates) to be redistributed to households and firms, notably those investing in renewable energy, 94 billion to be dedicated to social assistance for the unemployed and families, 87 billion of direct assistance to states to help them finance health expenditures of Medicaid, and finally 13 billion spent to improve access to digital technologies. The administration also attributed of 13. 4 billion dollars aid to automobile manufacturers General Motors and Chrysler, but this plan is not included in the stimulus plan.

These plans are meant to abate further economic contraction, however, with the present economic conditions differing from past recessions, in, that, many tenets of the American economy such as manufacturing, textiles, and technological development have been outsourced to other countries. Public works projects associated with the economic recovery plan outlined by the Obama Administration have been degraded by the lack of road and bridge development projects that were highly abundant in the Great Depression but are now mostly constructed and are mostly in need of maintenance. Regulations to establish market stability and confidence have been neglected in the Obama plan and have yet to be incorporated.

Federal Reserve Response

In an effort to increase available funds for commercial banks and lower the fed funds rate, on September 29 the U. S. Federal Reserve announced plans to double its Term Auction Facility to $300 billion. Because there appeared to be a shortage of U. S. dollars in Europe at that time, the Federal Reserve also announced it would increase its swap facilities with foreign central banks from $290 billion to $620 billion.

As of December 24, 2008, the Federal Reserve had used its independent authority to spend $1. 2 trillion on purchasing various financial assets and making emergency loans to address the financial crisis, above and beyond the $700 billion authorized by Congress from the federal budget. This includes emergency loans to banks, credit card companies, and general businesses, temporary swaps of treasury bills for mortgage-backed securities, the sale of Bear Stearns, and the bailouts of American International Group (AIG), Fannie Mae and Freddie Mac, and Citigroup.

Asia-Pacific Policy Responses

On September 15, 2008 China cut its interest rate for the first time since 2002. Indonesia reduced its overnight repo rate, at which commercial banks can borrow overnight funds from the central bank, by two percentage points to 10. 25 percent. The Reserve Bank of Australia injected nearly $1. 5 billion into the banking system, nearly three times as much as the market's estimated requirement. The Reserve Bank of India added almost $1. 32 billion, through a refinance operation, its biggest in at least a month. On November 9, 2008 the 2008 Chinese economic stimulus plan is a RMB¥ 4 trillion ($586 billion) stimulus package announced by the central government of the People's Republic of China in its biggest move to stop the global financial crisis from hitting the world's third largest economy. A statement on the government's website said the State Council had approved a plan to invest 4 trillion yuan ($586 billion) in infrastructure and social welfare by the end of 2010. The stimulus package will be invested in key areas such as housing, rural infrastructure, transportation, health and education, environment, industry, disaster rebuilding, income-building, tax cuts, and finance.

China's export driven economy is starting to feel the impact of the economic slowdown in the United States and Europe, and the government has already cut key interest rates three times in less than two months in a bid to spur economic expansion. On November 28, 2008, the Ministry of Finance of the People's Republic of China and the State Administration of Taxation jointly announced a rise in export tax rebate rates on some labour-intensive goods. These additional tax rebates will take place on December 1, 2008.

The stimulus package was welcomed by world leaders and analysts as larger than expected and a sign that by boosting its own economy, China is helping to stabilize the global economy. News of the announcement of the stimulus package sent markets up across the world. However, Marc Faber January 16 said that China according to him was in recession.

In Taiwan, the central bank on September 16, 2008 said it would cut its required reserve ratios for the first time in eight years. The central bank added $3. 59 billion into the foreign-currency interbank market the same day. Bank of Japan pumped $29. 3 billion into the financial system on September 17, 2008 and the Reserve Bank of Australia added $3. 45 billion the same day.

In developing and emerging economies, responses to the global crisis mainly consisted in low-rates monetary policy (Asia and the Middle East mainly) coupled with the depreciation of the currency against the dollar. There were also stimulus plans in some Asian countries, in the Middle East and in Argentina. In Asia, plans generally amounted to 1 to 3% of GDP, with the notable exception of China, which announced a plan accounting for 16% of GDP (6% of GDP per year).

European Policy Responses

Until September 2008, European policy measures were limited to a small number of countries (Spain and Italy). In both countries, the measures were dedicated to households (tax rebates) reform of the taxation system to support specific sectors such as housing. From September, as the financial crisis began to seriously affect the economy, many countries announced specific measures: Germany, Spain, Italy, Netherlands, United Kingdom, Sweden. The European Commission proposed a €200 billion stimulus plan to be implemented at the European level by the countries. At the beginning of 2009, the UK and Spain completed their initial plans, while Germany announced a new plan.

The European Central Bank injected $99. 8 billion in a one-day money-market auction. The Bank of England pumped in $36 billion. Altogether, central banks throughout the world added more than $200 billion from the beginning of the week to September 17.

On September 29, 2008 the Belgian, Luxembourg and Dutch authorities partially nationalized Fortis. The German government bailed out Hypo Real Estate. On 8 October 2008 the British Government announced a bank rescue package of around £500 billion ($850 billion at the time). The plan comprises three parts. First, £200 billion will be made available to the banks in the Bank of England's Special Liquidity scheme. Second, the Government will increase the banks' market capitalization, through the Bank Recapitalization Fund, with an initial £25 billion and another £25 billion to be provided if needed. Third, the Government will temporarily underwrite any eligible lending between British banks up to around £250 billion. In February 2009 Sir David Walker was appointed to lead a government inquiry into the corporate governance of banks.

In early December German Finance Minister Peer Steinbrück indicated that he does not believe in a "Great Rescue Plan" and indicated reluctance to spend more money addressing the crisis. In March 2009,

The European Union Presidency confirms that the EU is strongly resisting the US pressure to increase European budget deficits.

Global Responses

Most political responses to the economic and financial crisis has been taken, as seen above, by individual nations. Some coordination took place at the European level, but the need to cooperate at the global level has led leaders to activate the G-20 major economies entity. A first summit dedicated to the crisis took place, at the Heads of state level in November 2008 (2008 G-20 Washington summit).

The G-20 countries met in a summit held on November 2008 in Washington to address the economic crisis. Apart from proposals on international financial regulation, they pledged to take measures to support their economy and to coordinate them, and refused any resort to protectionism.

Another G-20 summit was held in London on April 2009. Finance ministers and central banks leaders of the G-20 met in Horsham on March to prepare the summit, and pledged to restore global growth as soon as possible. They decided to coordinate their actions and to stimulate demand and employment. They also pledged to fight against all forms of protectionism and to maintain trade and foreign investments. They also committed to maintain the supply of credit by providing more liquidity and recapitalizing the banking system, and to implement rapidly the stimulus plans. As for central bankers, they pledged to maintain low-rates policies as long as necessary. Finally, the leaders decided to help emerging and developing countries, through a strengthening of the IMF.

Countries Maintaining Growth or Technically Avoiding Recession

Poland is the only member of the European Union to have avoided a decline in GDP, meaning that in 2009 Poland has created the most GDP growth in the EU. As of December 2009 the Polish economy had not entered recession nor even contracted, while its IMF 2010 GDP growth forecast of 1. 9 per cent is expected to be upgraded. Analysts have identified several causes: Extremely low levels of bank lending and a relatively very small mortgage market; the relatively recent dismantling of EU trade barriers and the resulting surge in demand for Polish goods since 2004; the receipt of direct EU funding since 2004; lack of over-dependence on a single export sector; a tradition of government fiscal responsibility; a relatively large internal market; the free-floating Polish zloty; low labour costs attracting continued foreign direct investment; economic difficulties at the start of the decade which prompted austerity measures in advance of the world crisis; a government decision to refrain from quantative easing.

While China, India and Iran have experienced slowing growth, they have not entered recession.

South Korea narrowly avoided technical recession in the first quarter of 2009. The International Energy Agency stated in mid September that South Korea could be the only large OECD country to avoid recession for the whole of 2009. However, as of the October, the Australian economy has managed to avoid recession thanks largely to a strong mining sector and major stimulus spending, contracting only in the last quarter of 2008. It was the only developed economy to expand in the first half of 2009. On October 6, Australia became the first G20 country to raise its main interest rate, with the Reserve Bank of Australia deciding to move rates up to 3. 25% from 3. 00%. Australia has avoided a technical recession after experiencing only one quarter of negative growth in the fourth quarter of 2008, with GDP returning to positive in the first quarter of 2009.

Countries in Economic Recession or Depression

Many countries experienced recession in 2008. The countries/ territories currently in a technical recession are Estonia, Latvia, Ireland, New Zealand, Japan, Hong Kong, Singapore, Italy, Russia and Germany.

Denmark went into recession in the first quarter of 2008, but came out again in the second quarter. Iceland fell into an economic depression in 2008 following the collapse of its banking system. The following countries went into recession in the second quarter of 2008: Estonia, Latvia, Ireland and New Zealand.

The following countries/territories went into recession in the third quarter of 2008: Japan, Sweden, Hong Kong, Singapore, Italy, Turkey and Germany. As a whole the fifteen nations in the European Union that use the euro went into recession in the third quarter, and the United Kingdom. In addition, the European Union, the G7, and the OECD all experienced negative growth in the third quarter.

The following countries/territories went into technical recession in the fourth quarter of 2008: United States, Switzerland, Spain, and Taiwan.

South Korea "miraculously" avoided recession with GDP returning positive at a 0. 1% expansion in the first quarter of 2009.

Of the seven largest economies in the world by GDP, only China and France avoided a recession in 2008. France experienced a 0. 3% contraction in Q2 and 0. 1% growth in Q3 of 2008. In the year to the third quarter of 2008 China grew by 9%. This is interesting as China has until recently considered 8% GDP growth to be required simply to create enough jobs for rural people moving to urban centres. This figure may more accurately be considered to be 5–7% now that the main growth in working population is receding. Growth of between 5%–8% could well have the type of effect in China that a recession has elsewhere. Ukraine went into technical depression in January 2009 with a nominal annualized GDP growth of "20%. The recession in Japan intensified in the fourth quarter of 2008 with a nominal annualized GDP growth of "12. 7%, and deepened further in the first quarter of 2009 with a nominal annualized GDP growth of "15. 2%.

Official Forecasts in Parts of the World

On March 2009, U. S. Fed Chairman Ben Bernanke said in an interview that he felt that if banks began lending more freely, allowing the financial markets to return to normal, the recession could end during 2009. In that same interview, Bernanke said Green shoots of economic revival are already evident. On February 18, 2009, the US Federal Reserve cut their economic forecast of 2009, expecting the US output to shrink between 0. 5% and 1. 5%, down from its forecast in October 2008 of output between +1. 1% (growth) and "0. 2% (contraction).

The EU commission in Brussels updated their earlier predictions on January 19, 2009, expecting Germany to contract "2. 25% and "1. 8% on average for the 27 EU countries. According to new forecasts by Deutsche Bank (end of November 2008), the economy of Germany will contract by more than 4% in 2009.

On November 3, 2008, according to all newspapers, the European Commission in Brussels predicted for 2009 only an extremely low increase by 0. 1% of the GDP, for the countries of the Euro zone (France, Germany, Italy, etc.). They also predicted negative numbers for the UK ("1. 0%), Ireland, Spain, and other countries of the EU. Three days later, the IMF at Washington, D. C., predicted for 2009 a worldwide decrease, "0. 3%, of the same number, on average over the developed economies ("0. 7% for the US, and "0. 8% for Germany). On April 22, 2009, the German ministers of finance and that of economy, in a common press conference, corrected again their numbers for 2009 downwards: this time the "prognosis" for Germany was a decrease of the GDP of at least "5%, in agreement with a recent prediction of the IMF.

On June 11, 2009, the World Bank Group predicted for 2009 for the first time a global contraction of the economic power, precisely by "3%.

Comparisons with the Great Depression

Although some casual comparisons between the late-2000s recession and the Great Depression have been made, there remain large differences between the two events. The consensus among economists in March 2009 was that a depression was not likely to occur. UCLA Anderson Forecast director Edward Leamer said on March 25, 2009 that there had not been any major predictions at that time which resembled a second Great Depression:

> *"We've frightened consumers to the point where they imagine there is a good prospect of a Great Depression. That certainly is not in the prospect. No reputable forecaster is producing anything like a Great Depression."*

Differences explicitly pointed out between the recession and the Great Depression include the facts that over the 79 years between 1929 and 2008, great changes occurred in economic philosophy and policy, the stock market had not fallen as far as it did in 1932 or 1982, the 10-year price-to-earnings ratio of stocks was not as low as in the '30s or '80s, inflation-adjusted U. S. housing prices in March 2009 were higher than any time since 1890 (including the housing booms of the 1970s and '80s), the recession of the early '30s lasted over three-and-a-half years, and during the 1930s the supply of money (currency plus demand deposits) fell by 25% (where as in 2008 and 2009 the Fed "has taken an ultraloose credit stance"). Furthermore, the unemployment rate in 2008 and early 2009 and the rate at which it rose was comparable to most of the recessions occurring after World War II, and was dwarfed by the 25% unemployment rate peak of the Great Depression.

Price-to-earnings ratios have yet to drop as low as in previous recessions. On this issue, "it is critically important, though, to recognize that different analysts have different earnings expectations, and the consensus view is more often wrong than right." Some argue that price-to-earnings ratios remain high because of unprecedented falls in earnings.

Three years into the Great Depression, unemployment reached a peak of 25% in the U. S. The United States entered into recession in December 2007 and in March 2009, U-3 unemployment reached 8. 5%. In March 2009, statistician John Williams "argue[d] that measurement changes implemented over the years make it impossible to compare the current unemployment rate with that seen during the Great Depression".

Nobel Prize winning Economist Paul Krugman predicted a series of depressions in his *Return to Depression Economics* (2000), based on "failures on the demand side of the economy." On January 5, 2009, he wrote that "preventing depressions isn't that easy after all" and that "the economy is still in free fall." In March 2009, Krugman explained that a major difference in this situation is that the causes of this financial crisis were from the shadow banking system. "The crisis hasn't involved problems with deregulated institutions that took new risks... Instead, it involved risks taken by institutions that were never regulated in the first place." On February 22, NYU economics professor Nouriel Roubini said that the crisis was the worst since the Great Depression, and that without cooperation between political parties and foreign countries, and if poor fiscal policy decisions (such as support of zombie banks) are pursued, the situation "could become as bad as the Great Depression." On April 27, 2009, Roubini expressed a more upbeat assessment by noting that "the bottom of the economy [will be seen] toward the beginning or middle of next year."

Market strategist Phil Dow "said he believes distinctions exist between the current market malaise" and the Great Depression. The Dow's fall of over 50% in 17 months is similar to a 54. 7% fall in the Great Depression, followed by a total drop of 89% over the next 16 months. "It's very troubling if you have a mirror image," said Dow. Floyd Norris, chief financial correspondent of The New York Times, wrote in a blog entry in March 2009 that the decline has not been a mirror image of the Great Depression, explaining that although the decline amounts were nearly the same at the time, the rates of decline had started much faster in 2007, and that the past year had only ranked eighth among the worst recorded years of percentage drops in the Dow. The past two years ranked third however.

On November 15, 2008, author and SMU economics professor Ravi Batra said he is "afraid the global financial debacle will turn into a steep recession and be the worst since the Great Depression, even worse than the painful slump of 1980–1982 that afflicted the whole world". In 1978, Batra's book *The Downfall of Capitalism and Communism* was published. His first major prediction came true with the collapse of Soviet Communism in 1990. His second major prediction for a financial crisis to engulf the capitalist system seems to be unfolding since 2007 with increasing attention being paid to his work. In his final press conference as president, George W. Bush claimed that in September 2008 his chief economic advisors had said that the economic situation could at some point become worse than the Great Depression.

A tent city in Sacramento, California was described as "images, hauntingly reminiscent of the iconic photos of the 1930s and the Great Depression" and "evocative Depression-era images."

On April 6, 2009 Vernon L. Smith and Steven Gjerstad offered the hypothesis "that a financial crisis that originates in consumer debt, especially consumer debt concentrated at the low end of the wealth and income distribution, can be transmitted quickly and forcefully into the financial system. It appears that we're witnessing the second great consumer debt crash, the end of a massive consumption binge."

On April 17, 2009, head of the IMF Dominique Strauss-Kahn said that there was a chance that certain countries may not implement the proper policies to avoid feedback mechanisms that could eventually turn the recession into a depression. "The free-fall in the global economy may be starting to abate, with a recovery emerging in 2010, but this depends crucially on the right policies being adopted today." The IMF pointed out that unlike the Great Depression, this recession was synchronized by global integration of markets. Such synchronized recessions were explained to last longer than typical economic downturns and have slower recoveries.

In South Africa

On February 11, South Africa's Finance Minister Trevor Manuel said that "what started as a financial crisis might well become a second Great Depression."

In the United Kingdom

On February 10, 2009, Ed Balls, Secretary of State for Children, Schools and Families of the United Kingdom, said that "I think that this is a financial crisis more extreme and more serious than that of the 1930s and we all remember how the politics of that era were shaped by the economy." On January 24, 2009, Edmund Conway, Economics Editor for *The Daily Telegraph*, wrote that "The plight facing Britain is uncannily similar to the 1930s, since prices of many assets – from shares to house prices – are falling at record rates [in Britain], but the value of the debt against which they are held remains unchanged."

In Ireland

The Republic of Ireland "technically" entered into an economic depression in 2009. The ESRI (Economic and Social Research Institute) predict an economic contraction of 14% by 2010, however this number may have already been exceeded with GDP dropping 7. 1% quarter on quarter during the fourth quarter of 2008, and a possible greater contraction in the first quarter of 2009 with the fall in all OECD countries with the exception of France exceeding the drop of the previous quarter. Unemployment is up 8. 75% to 11. 4%. Government borrowing and the financial bailout and Nationalisation of one of Ireland's banks which were loaded with debt due to the Irish property bubble.

Job losses and Unemployment Rates

Many jobs have been lost worldwide. In the US, job loss has been going on since December 2007, and it accelerated drastically starting in September 2008 following the bankruptcy of Lehman Brothers. By February 2010, the American economy is reported to be more shaky than the economy of Canada. Many service industries have reported dropping their prices in order to maximize profit margins in an era where employment is fleeting and either underemployment on unemployment is a sure factor in life.

Net job gains and losses by month in the United States;

- September 2008 – 280,000 jobs lost
- October 2008 – 240,000 jobs lost
- November 2008 – 333,000 jobs lost
- December 2008 – 632,000 jobs lost
- January 2009 – 741,000 jobs lost
- February 2009 – 681,000 jobs lost
- March 2009 – 652,000 jobs lost
- April 2009 – 519,000 jobs lost
- May 2009 – 303,000 jobs lost
- June 2009 – 463,000 jobs lost
- July 2009 – 276,000 jobs lost
- August 2009 – 201,000 jobs lost
- September 2009 – 263,000 jobs lost
- October 2009 – 111,000 jobs lost
- November 2009-64,000 jobs created
- December 2009-109,000 jobs lost
- January 2010-14,000 jobs created
- February 2010-39,000 jobs created
- March 2010-208,000 jobs created
- April 2010-290,000 jobs created

- May 2010-413,000 jobs created
- June 2010-125,000 jobs lost
- 2008 (September 2008 – December 2008) – 2. 6 million jobs lost
- 2009 (January 2009 – December 2009) – 4. 2 million jobs lost
- 2010 (January 2010–present)-n/a
- Current unemployment rate: 9. 5%.

Since the start of 2008, 6. 7 million jobs have been lost, according to the Bureau of Labour Statistics.

Canada Net Job Losses and Gains by Month

Drastic job loss in Canada started later than in the US. Some months in 2008 had job growth, such as September, while others such as July had losses. Due to the collapse of the American car industry at the same time as a strong Canadian dollar achieved parity +10% against a poorly-performing US dollar, the cross-border manufacturing industry has been disproportionately affected throughout.

- September 2008 – No net loss
- October 2008 – No net loss
- November 2008 – 71,000 jobs lost
- December 2008 – 34,000 jobs lost
- January 2009 – 129,000 jobs lost
- February 2009 – 82,600 jobs lost
- March 2009 – 61,300 jobs lost
- April 2009 – No net loss (1)
- May 2009 – 36,000 jobs lost
- October 2009 – 43,200 jobs lost
- December 2009-28,300 jobs lost
- January 2010-43,000 jobs created
- February 2010-21,000 jobs created
- March 2010-17,900 jobs created
- April 2010-109,000 jobs created
- May 2010-25,000 jobs created
- June 2010-20,000 jobs created.

37,000 jobs are gained in the self-employment category: While job creation has increased in the past three months, most Canadians still complain about people getting laid off from work. However, a growing amount of these layoffs are considered for seasonal purposes and has little or no relation to the recession. Excluding Stelco employees, most laid off workers have six months to acquire a job while collecting unemployment insurance. After that, they must go on welfare and continue their job search from there.

- May 2009 Canadian unemployment rate: 8. 4%
- September 2009 Canadian unemployment rate: 8. 7%
- November 2009 Canadian unemployment rate: 8. 6%
- January 2010 Canadian unemployment rate: 8. 3%
- May 2010 Canadian unemployment rate: 8. 1%
- June 2010 Canadian unemployment rate: 8. 1%.

The unemployment rate has been stabilized between 8. 0% and 11. 0% for the past two years; signifying the economic strength of Canada's financial institutions compared to its counterparts in the United States. Many job places in Canada have opted to reduce hours rather than lay off staff-especially in industries that are needed to keep the economy from going into a depression. While the automotive sector is slowly recalling workers back to work, grocery stores and restaurants have slashed hours in the face of the global recession.

Australia net job losses and gains by month;

- September 2008 – 2,200 jobs created
- October 2008 – 34,300 jobs created
- November 2008 – 15,600 jobs lost
- December 2008 – 1,200 jobs lost
- January 2009 – 1,200 jobs created
- February 2009 – 1,800 jobs created
- March 2009 – 34,700 jobs lost
- April 2009 – 27,300 jobs created
- May 2009 – 1,700 jobs lost
- June 2009 – 21,400 jobs lost
- July 2009 – 32,200 jobs created
- August 2009 – 27,100 jobs lost
- September 2009 – 40,600 jobs created
- October 2009 – 24,500 jobs created.
    * April 2009 Australian unemployment rate: 5. 5%
    * July 2009 Australian unemployment rate: 5. 8%
    * August 2009 Australian unemployment rate: 5. 8%

* September 2009 Australian unemployment rate: 5. 7%
* October 2009 Australian unemployment rate: 5. 8%.

The unemployment rate for October rose slightly due to population growth and other factors leading to 35,000 people looking for work, even though 24,500 jobs were created. In general, throughout the subdued economic growth caused by the recession in the rest of the world, Australian employers have elected to cut working hours rather than fire employees, in recognition of the skill shortage caused by the resources boom.

Oil Price Increases Since 2003

From the mid-1980s to September 2003, the inflation-adjusted price of a barrel of crude oil on NYMEX was generally under $25/barrel. During 2003, the price rose above $30, reached $60 by August 11, 2005, and peaked at $147. 30 in July 2008. Commentators attributed these price increases to many factors, including reports from the United States Department of Energy and others showing a decline in petroleum reserves, worries over peak oil, Middle East tension, and oil price speculation.

For a time, geo-political events and natural disasters indirectly related to the global oil market had strong short-term effects on oil prices, such as North Korean missile tests, the 2006 conflict between Israel and Lebanon, worries over Iranian nuclear plans in 2006, Hurricane Katrina, and various other factors.

By 2008, such pressures appeared to have a insignificant impact on oil prices given the onset of the global recession. The recession caused demand for energy to shrink in late 2008, with oil prices falling from the July 2008 high of $147 to a December 2008 low of $32. Oil prices stabilized by October 2009 and established a trading range between $60 and $80.

New Inflation-adjusted Records

The price of crude oil in 2003 traded in a range between $20–$30/ bbl. Between 2003 and July 2008, prices steadily rose, reaching $100/ bbl in late 2007, tying the previous all time inflation-adjusted record set in 1980. A steep rise in the price of oil in 2008-also mirrored by other commodities-culminated in an all time high of $147. 27 during trading on July 11, 2008, more than a third above the previous inflation-adjusted high.

High oil prices and economic weakness contributed to a demand contraction in 2007-2008. In the United States, gasoline consumption declined by 0. 4% in 2007, then fell by 0. 5% in the first two months of

2008 alone. Record-setting oil prices in the first half of 2008 and economic weakness in the second half of the year prompted a 1. 2 million bbl/day contraction in US consumption of petroleum products, representing 5. 5% of total US consumption, the largest decline since 1980 at the climax of the 1979 energy crisis.

Possible Causes

Demand: World crude oil demand grew an average of 1. 76% per year from 1994 to 2006, with a high of 3. 4% in 2003-2004. World demand for oil is projected to increase 37% over 2006 levels by 2030, according to the 2007 U. S. Energy Information Administration's (EIA) annual report. In 2007, the EIA expected demand to reach an ultimate high of 118 million barrels per day (18. 8×10^{6}△ m/d), from 2006's 86 million barrels (13. 7×10△^{6} m^{3}), driven in large part by the transportation sector.

A 2008 report from the International Energy Agency (IEA) predicted that although drops in petroleum demand due to high prices have been observed in developed countries and are expected to continue, a 3. 7 percent rise in demand by 2013 is predicted in developing countries. This is projected to cause a net rise in global petroleum demand during that period.

Transportation consumes the largest proportion of energy, and has seen the largest growth in demand in recent decades. This growth has largely come from new demand for personal-use vehicles powered by internal combustion engines.

This sector also has the highest consumption rates, accounting for approximately 68. 9% of the oil used in the United States in 2006, and 55% of oil use worldwide as documented in the Hirsch report. Cars and trucks are predicted to cause almost 75% of the increase in oil consumption by India and China between 2001 and 2025. In 2008, auto sales in China have been expected to grow by as much as 15-20 percent, resulting in part from economic growth rates of over 10 percent for 5 years in a row.

Demand growth is highest in the developing world, but the United States is the world's largest consumer of petroleum. Between 1995 and 2005, US consumption grew from 17. 7 million barrels a day to 20. 7 million barrels a day, a 3 million barrel a day increase. China, by comparison, increased consumption from 3. 4 million barrels a day to 7 million barrels a day, an increase of 3. 6 million barrels a day, in the same time frame. Per capita, annual consumption by people in the US is 24. 85 barrels, 1. 79 barrels in China, and. 79 barrels in India.

As countries develop, industry, rapid urbanization and higher living standards drive up energy use, most often of oil. Thriving economies such as China and India are quickly becoming large oil consumers.

China has seen oil consumption grow by 8% yearly since 2002, doubling from 1996-2006. In 2008, auto sales in China were expected to grow by as much as 15-20 percent, resulting in part from economic growth rates of over 10 percent for 5 years in a row. Although swift continued growth in China is often predicted, others predict that China's export dominated economy will not continue such growth trends due to wage and price inflation and reduced demand from the US. India's oil imports are expected to more than triple from 2005 levels by 2020, rising to 5 million barrels per day (790×10△³ m³/d).

Another large factor on petroleum demand has been human population growth. Because world population grew faster than oil production, production per capita peaked in 1979 (preceded by a plateau during the period of 1973-1979). The world's population in 2030 is expected to be double that of 1980.

The Role of Fuel Subsidies

State fuel subsidies have shielded consumers in many nations from the price rises, but many of these subsidies are being reduced or removed as the cost to governments of subsidization increases.

In June 2008, AFP reported that:

> *"China became the latest Asian nation to curb energy subsidies last week after hiking retail petrol and diesel prices as much as 18 percent... Elsewhere in Asia, Malaysia has hiked fuel prices by 41 percent and Indonesia by around 29 percent, while Taiwan and India have also raised their energy costs."*

In the same month, Reuters reported that:

> *"Countries like China and India, along with Gulf nations whose retail oil prices are kept below global prices, contributed 61 percent of the increase in global consumption of crude oil from 2000 to 2006, according to JPMorgan.*

Other than Japan, Hong Kong, Singapore and South Korea, most Asian nations subsidize domestic fuel prices. The more countries subsidize them, the less likely high oil prices will have any affect [*sic*] in reducing overall demand, forcing governments in weaker financial situations to surrender first and stop their subsidies.

That is what happened over the past two weeks. Indonesia, Taiwan, Sri Lanka, Bangladesh, India, and Malaysia have either raised regulated fuel prices or pledged that they will."

The Economist reported: "Half of the world's population enjoys fuel subsidies.

This estimate, from Morgan Stanley, implies that almost a quarter of the world's petrol is sold at less than the market price." U. S. Secretary of Energy Samuel Bodman stated that around 30 million barrels per day (4,800,000 m^3/d) of oil consumption (over a third of the global total) is subsidized. But energy analyst Jeff Vail warned that cutting subsidies would do little to reduce global prices.

Supply

An important contributor to price increases has been the slow down in oil supply growth, which has continued since oil production surpassed new discoveries in 1980. The fact that global oil production will decline at some point, leading to lower supply is the main long-term fundamental cause of rising prices.

Although there is contention about the exact time at which global production will peak, there are now very few parties who do not acknowledge that the concept of a production peak is valid. However, before the record oil prices of 2008, some commentators argued that global warming awareness and new energy sources would limit demand before the effects of supply could, suggesting that reserve depletion would be a non-issue.

A large factor in the lower supply growth of petroleum has been that oil's historically high ratio of Energy Returned on Energy Invested is in significant decline. Petroleum is a limited resource, and the remaining accessible reserves are consumed more rapidly each year. Remaining reserves are increasingly more technically difficult to extract and therefore more expensive.

Eventually, reserves will only be economically feasible to extract at extremely high prices. Even if total oil supply does not decline, increasing numbers of experts believe the easily accessible sources of light sweet crude are almost exhausted and in the future the world will depend on more expensive unconventional oil reserves and heavy oil, as well as renewable energy sources. It is thought by many, including energy economists such as Matthew Simmons, that prices could continue to rise indefinitely until a new market equilibrium is reached at which point supply satisfies worldwide demand.

A prominent example of investment in non-conventional sources is seen in the Canadian tar sands. They are a far less cost-efficient source of heavy, low-grade oil than conventional crude, but when oil trades above $60/bbl, the tar sands become attractive to exploration and production companies. While Canada's tar sands region is estimated to contain as much "heavy" oil as all the world's reserves of "conventional" oil, efforts to economically exploit these resources lag behind the increasing demand of recent years.

Until 2008, CERA (a consulting company wholly owned by energy consultants IHS Energy) did not believe this would be such an immediate problem.

However, in an interview with *The Wall Street Journal,* Daniel Yergin, previously known for his quotes that the price of oil would soon return down to "normal", amended the company's position on May 7, 2008 to predict that oil would reach $150 during 2008, due to tightness of supply This reversal of opinion was significant, as CERA, among other consultancies, provided price projections that were used by many official bodies to plan long term strategy in respect of energy mix and price.

Other major energy organisations, such as the International Energy Agency (IEA), had already been much less optimistic in their assessments for some time. In 2008, the IEA drastically reduced its prediction of production decline from 3. 7% a year to 6. 7% a year, based largely on better accounting methods, including actual research of individual oil field production through out the world.

Terrorist and insurgent groups have increasingly targeted oil and gas installations. Sometimes, such attacks are perpetrated by militias in regions where oil wealth has produced few tangible benefits for the local citizenry, as is the case in the Niger Delta.

Many factors have resulted in possible and/or actual concerns about reduced supply of oil. The post-9/11 war on terror, Labour strikes, hurricane threats to oil platforms, fires and terrorist threats at refineries, and other short-lived problems are not solely responsible for the higher prices. Such problems do push prices higher temporarily, but have not historically been fundamental to long-term price increases.

Investment Demand

Investment demand for oil occurs when investors purchase *futures contracts* to buy a commodity at a set price for future delivery. "Speculators are not buying any actual crude.... When [the] contracts mature, they either settle them with a cash payment or sell them on to genuine consumers."

Several claims have been made implicating financial speculation as a major cause of the price increases. In May 2008 the transport chief for Germany's Social Democrats estimated that 25 percent of the rise to $135 a barrel had nothing to do with underlying supply and demand.

Testimony was given to a U. S. Senate committee in May indicating that "demand shock" from "Institutional Investors" had increased by 848 million barrels (134,800,000 m) over the last five years, similar to increases in demand from China (920 million barrels). The influence of Institutional Investors, such as sovereign-wealth funds, was also discussed in June 2008, when Lehman Brothers suggested that price increases were related to increases in exposure to commodities by such investors. It claimed that "for every $100 million in new inflows, the price of West Texas Intermediate, the U. S. benchmark, increased by 1. 6%." Also in May 2008, an article in *The Economist* pointed out that oil futures transactions on the New York Mercantile Exchange (NYMEX), nearly mirrored the price of oil increases for a several year period, however the article conceded that the increased investment might be following rising prices, rather than causing them, and that the nickel commodity market had halved in value between May 2007 and May 2008 despite significant speculative interest.

It also reminds readers "Investment can flood into the oil market without driving up prices because speculators are not buying any actual crude... no oil is hoarded or somehow kept off the market," and that prices of some commodities which are not openly traded have actually risen faster than oil prices.

In June 2008, OPEC's Secretary General Abdullah al-Badri stated that current world consumption of oil at 87 million bpd was far exceeded by the "paper market" for oil, which equals about 1. 36 billion bpd, or more than 15 times the actual market demand.

In response to the possibility that financial speculators artificially inflated the oil market, the U. S. Congress began hearings in June 2008 to discover if actions to "tighten restrictions on pension funds, investment banks and other investors that they say are driving up fuel prices" were necessary.

An interagency task force on commodities markets was formed in the U. S. government to investigate the claims of speculators influence on the petroleum market concluded in July 2008 that "market fundamentals" such as supply and demand provide the best explanations for oil price increases, and that increased speculation was not statistically correlated with the increases. The report also noted that increased prices with an elastic supply would cause increases in petroleum inventories. As inventories have actually declined, the task force concluded market pressures are most likely to blame. Similarly, other commodities which are not subject to market speculation (such as coal, steel, and onions) have seen similar price increases over the same time period.

In June 2008 U. S. energy secretary Samuel Bodman had said that insufficient oil production, not financial speculation, was driving rising crude prices. He said that oil production has not kept pace with growing demand. "In the absence of any additional crude supply, for every 1% of crude demand, we will expect a 20% increase in price in order to balance the market," Bodman said. This contradicts earlier statements by Iranian OPEC governor Mohammad-Ali Khatibi indicating that the oil market is saturated and that an increase in production announced by Saudi Arabia was "wrong". OPEC itself had also previously stated that the oil market was well supplied and that high prices were a result of speculation and a weak U. S. dollar.

In September 2008, a study of the oil market by Masters Capital Management was released which claimed that speculation did significantly impact the market. The study stated that over $60 billion was invested in oil during the first 6 months of 2008, helping drive the price per barrel from $95 to $147 per barrel, and that by the beginning of September, $39 billion had been withdrawn by speculators, causing prices to fall.

Monetary Inflation and the Value of the US Dollar

The Austrian School of economics holds that price inflation derives from monetary inflation, and its advocates, such as the Ludwig von Mises Institute and congressman Ron Paul, argue that loose monetary policy from the Federal Reserve and other central banks is a major contributor to the increase in oil prices, and the cause of both commodity speculation and dollar devaluation.

Oil is quoted and traded in US Dollars. Therefore by definition the price or value of oil fluctuates based on investors' assessment of the value of oil and US Dollars. This has led to concern among some economists that the principal earned from the sale of oil may lose value in the long run if the U. S. dollar loses real value. Some analysts believe that as much as $25 of the June 2008 prices around $140 were due to dollar devaluation.

The US Dollar price of oil demonstrated a strong positive correlation with the EUR/USD exchange rate from 2006 to 2010.

Effects

There is debate over what the effects of the 2000s energy crisis will be over the long term. Some speculate that an oil-price spike could create a recession comparable to those that followed the 1973 and 1979 energy crises or a potentially worse situation such as a global oil crash.

Increased petroleum prices are however reflected in a vast number of products derived from petroleum, as well as those transported using petroleum fuels. Political scientist George Friedman has postulated that if high prices for oil and food persist, they will define the fourth distinct geopolitical regime since the end of World War II, the previous three being the Cold War, the 1989-2001 period in which economic globalization was primary, and the post-9/11 "war on terror". In addition to high oil prices, from year 2000 volatility in the price of oil has increased notably and this volatility has been suggested to be a factor in the ongoing financial crisis which began in 2008.

Forecasted Prices and Trends

According to informed observers, OPEC, meeting in early December, 2007, seemed to desire a high but stable price that would deliver substantial needed income to the oil producing states, but avoid prices so high that they would negatively impact the economies of the oil consuming nations. A range of 70–80 dollars a barrel was suggested by some analysts to be OPEC's goal. Some analysts point out that major oil exporting countries are rapidly developing; and because they are using more oil domestically, less oil may be available on the international market. This effect, outlined in the export land economic model, could significantly reduce the oil available for trade and cause prices to continue to rise. Particularly significant are Indonesia (which is now a net importer of oil), Mexico and Iran (where demand is projected to exceed production in about 5 years), and Russia (whose domestic petroleum demand is growing rapidly).

In May 2008, T. Boone Pickens, an influential oil investor who believes the world's oil output is about to peak, predicted oil prices would hit $150 a barrel by the end of the year. "Eighty-five million barrels of oil a day is all the world can produce, and the demand is 87 m," Mr. Pickens said in an interview with CNBC. "It's just that simple."

In June 2008, Alexei Miller, head of Russian energy giant Gazprom, warned that the price of oil is likely to hit $250 a barrel sometime in 2009. Miller said that while speculation had played a role in oil prices, "this influence was not decisive." Bloomberg reported that, as of mid-

June, "At least 3,008 options contracts have been purchased giving holders the right to buy oil at $250 a barrel in December". Also in June 2008, Shukri Ghanem, head of Libya's National Oil Corporation, said: "I think it [the oil price] will go higher. That is a trend that will continue for some time. The easy, cheap oil is over, peak oil is looming."

On June 26, 2008, OPEC President Chakib Khelil said in an interview: "I forecast prices probably between $150-170 during this summer. That will perhaps ease towards the end of the year." Iran's OPEC governor Mohammad-Ali Khatibi predicts that the price of oil would reach $150 a barrel by the end of this summer.

Near-term peak oil proponent Matthew Simmons predicts a rise to $300 a barrel or higher by 2013 as sweet crude petroleum becomes more scarce and major producers begin failing to meet demand.

In November, as prices fell below $60 a barrel, the IEA warned that falling prices may create both a lack of investment in new sources of oil and a fall in production of more expensive unconventional reserves such as the tar sands of Canada. The IEA's chief economist warned, "Oil supplies in the future will come more and more from smaller and more difficult fields," meaning that future production requires more investment every year.

A lack of new investment in such projects, which had already been observed, could eventually cause new and more severe supply issues than had been experienced in the early 2000s according to the IEA. Because the sharpest production declines had been seen in developed countries, the IEA warned that the greatest growth in production was expected to come from smaller projects in OPEC states, raising their world production share from 44% in 2008 to a projected 51% in 2030. The IEA also pointed out that demand from the developed world may have also peaked, so that future demand growth was likely to come from developing nations such as China, contributing 43%, and India and the Middle East, each about 20%).

Timothy Kailing argued against the IEA's earlier predictions in a 2008 Journal of Energy Security article. He pointed out the difficulty of increasing production even with vastly increased investment in exploration and production in mature petroleum regions.

By looking at the historical response of production to variation in drilling effort, this analysis claimed that very little increase of production could be attributed to increased drilling. This was due to a tight the quantitative relationship of diminishing returns with increasing drilling effort: as drilling effort increased, the energy obtained per active drill rig was reduced according to a severely diminishing power law. This fact means that even an enormous increase of drilling effort is unlikely to lead to significantly increased oil and gas production in a mature petroleum region like the United States.

In its 2008 World Energy Outlook, the International Energy Agency (IEA) predicted a rate of decline in output from the world's existing oilfields of 6. 7% a year.

The End of the Crisis

By the beginning of September 2008, prices had fallen to $110. OPEC secretary Abdalla El-Badri said that it intended to cut output by about 500,000 barrels a day, which he saw as correcting a "huge oversupply" due to declining economies and a stronger U. S. dollar. On September 10, the International Energy Agency (IEA) lowered its 2009 demand forecast by 140,000 barrels to 87. 6 million barrels a day.

As many countries throughout the world entered economic recession in the third quarter of 2008, prices continued to slide. In November and December, global demand growth fell, and U. S. demand fell 10% overall from early October to early November 2008 (accompanying a significant drop in auto sales). In their December meeting, OPEC planned to reduce their production by 2. 2 million barrels per day, though they admitted their resolution to reduce production in October had only an 85% compliance rate. Petroleum prices had fallen to below $35 in February 2009, but on May 6, 2009 had risen back to mid-November 2008 levels at about $56. The global economic downturn left oil storage facilities with more oil than in any year since 1990, when Iraq's invasion of Kuwait upset the market.

Possible Mitigations

Attempts to mitigate the impacts of oil price increases include:

- Increasing the supply of petroleum
- Finding substitutes for petroleum
- Decreasing the demand for petroleum
- Attempting to reduce the impact of rising prices on petroleum consumers.

In mainstream economic theory, a free market rations an increasingly scarce commodity by increasing its price. A higher price should stimulate producers to produce more, and consumers to consume less, while possibly shifting to substitutes. The first three mitigation strategies in the above list are, therefore, in keeping with mainstream economic theory, as government policies can affect the supply and demand for petroleum as well as the availability of substitutes. In contrast, the last type of strategy in the list (attempting to shield consumers from rising prices) would seem to work against classical economic theory, by encouraging consumers to overconsume the scarce quantity, thus making it even scarcer. To avoid creating outright shortages, attempts at price control may require some sort of rationing scheme.

Alternative Propulsion

Alternative Fuels

Economists say that the substitution effect will spur demand for alternate fossil fuels, such as coal or liquefied natural gas and for renewable energies, such as solar power, wind power, and advanced biofuels. For example, China and India are currently heavily investing in natural gas and coal liquefaction facilities. Nigeria is working on burning natural gas to produce electricity instead of simply flaring the gas, where all non-emergency gas flaring will be forbidden after 2008. Outside the U. S., more than 50% of oil is consumed for stationary, non-transportation purposes such as electricity production where it is relatively easy to substitute natural gas for oil.

Ironically, oil companies including the supermajors have begun to fund research into alternative fuel. BP has invested half a billion dollars for research over the next several years. The motivations behind such moves are to acquire the patent rights as well as understanding the technology so vertical integration of the future industry could be achieved.

Bioplastics and Bioasphalt

Another major factor in petroleum demand is the widespread use of petroleum products such as plastic. These could be partially replaced by bioplastics, which are derived from renewable plant feedstocks such as vegetable oil, cornstarch, pea starch, or microbiota.

They are used either as a direct replacement for traditional plastics or as blends with traditional plastics. The most common end use market is for packaging materials. Japan has also been a pioneer in bioplastics, incorporating them into electronics and automobiles.

United States Strategic Fuel Reserve

The United States Strategic Petroleum Reserve could, on its own, supply current U. S. demand for about a month in the event of an emergency, unless it were also destroyed or inaccessible in the emergency.

This could potentially be the case if a major storm were to hit the Gulf of Mexico, where the reserve is located. While total consumption has increased, the western economies are less reliant on oil than they were twenty-five years ago, due both to substantial growth in productivity and the growth of sectors of the economy with little oil dependence such as finance and banking, retail, etc.

The decline of heavy industry and manufacturing in most developed countries has reduced the amount of oil per unit GDP; however, since these items are imported anyway, there is less change in the oil dependence of industrialized countries than the direct consumption statistics indicate.

Fuel Taxes

One recourse used and discussed in the past to avoid the negative impacts of oil shocks in the many developed countries which have high fuel taxes has been to temporarily or permanently suspend these taxes as fuel costs rise. France, Italy, and the Netherlands lowered taxes in 2000 in response to protests over high prices, but other European nations resisted this option because public service financiation is partly based on energy taxes.

The issue came up again in 2004, when oil reached $40 a barrel causing a meeting of 25 EU finance ministers to lower economic growth forecasts for that year. Because of budget deficits in several countries, they decided to pressure OPEC to lower prices instead of lowering taxes. In 2007, European truckers, farmers, and fishermen again raised concerns over record oil prices cutting into their earnings, hoping to have taxes lowered. In the United Kingdom, where fuel taxes were raised in October and are scheduled to rise again in April 2008, there was talk of protests and roadblocks if the tax issue was not addressed. On April 1, 2008, a 25 yen per liter fuel tax in Japan was allowed to lapse temporarily.

This method of softening price shocks is even less viable to countries with much lower gas taxes, such as the United States.

Demand Management

Transportation demand management has the potential to be an effective policy response to fuel shortages or price increases and has a greater probability of long term benefits than other mitigation options.

There are major differences in energy consumption for private transport between cities; an average U. S. urban dweller uses 24 times more energy annually for private transport as a Chinese urban resident.

These differences cannot be explained by wealth alone but are closely linked to the rates of walking, cycling, and public transport use and to enduring features of the city including urban density and urban design. For individuals, telecommuting provides alternatives to daily commuting and long-distance air travel for business. Technologies for telecommuting, such as video conferencing, e-mail, and corporate wikis, continue to improve, in keeping with the overall improvement in information technologies ascribed to Moore's law.

As the cost of moving information by moving human workers continues to rise, while the cost of moving information electronically continues to fall, presumably market forces should cause more people to substitute virtual travel for physical travel. Matthew Simmons explicitly calls for "liberating the workforce" by changing the corporate mindset from paying people to show up physically to work every day, to paying them instead for the work they do, from any location.

This would allow many more information workers to work from home either part-time or full-time, or from satellite offices or Internet cafes near to where they live, freeing them from long daily commutes to central offices.

High energy prices and a slowed economy caused petroleum consumption to reach a three year low in crude oil imports to the United States in December 2008.

Political Action Against Market Speculation

The price rises of mid-2008 led to a variety of proposals to change the rules governing energy markets and energy futures markets, in order to prevent rises due to market speculation.

On July 26, 2008, the United States House of Representatives passed the Energy Markets Emergency Act of 2008 (H. R. 6377), which directs the Commodity Futures Trading Commission (CFTC) "to utilize all its authority, including its emergency powers, to curb immediately the role of excessive speculation in any contract market within the jurisdiction and control of the Commodity Futures Trading Commission, on or through which energy futures or swaps are traded, and to eliminate excessive speculation, price distortion, sudden or unreasonable fluctuations or unwarranted changes in prices, or other unlawful activity causing major market disturbances that prevent the market from accurately reflecting the forces of supply and demand for energy commodities.

Chapter 6

Overview of World Geography

Asia is the world's largest and most populous continent, located primarily in the eastern and northern hemispheres. It covers 8.6% of the Earth's total surface area (or 29.9% of its land area) and with approximately 4 billion people, it hosts 60% of the world's current human population. During the 20th century Asia's population nearly quadrupled.

Asia is traditionally defined as part of the landmass of Eurasia — with the western portion of the latter occupied by Europe — located to the east of the Suez Canal, east of the Ural Mountains and south of the Caucasus Mountains (or the Kuma-Manych Depression) and the Caspian and Black Seas. It is bounded on the east by the Pacific Ocean, on the south by the Indian Ocean and on the north by the Arctic Ocean. Given its size and diversity, Asia — a toponym dating back to classical antiquity — is more a cultural concept incorporating a number of regions and peoples than a homogeneous physical entity.

The wealth of Asia differs very widely among and within its regions, due to its vast size and huge range of different cultures, environments, historical ties and government systems. In terms of nominal GDP, Japan has the largest economy on the continent and the second largest in the world. In purchasing power parity terms, however, China has the largest economy in Asia and the second largest in the world.

Etymology

The term "Asia" is originally a concept exclusively of Western civilization. The peoples of ancient *Asia* (Chinese, Japanese, Indians, Persians, Arabs etc.) never conceived the idea of *Asia*, simply because they did not see themselves collectively. In their perspective, they were vastly varied civilizations, contrary to ancient European belief.

The word *Asia* originated from the Greek word *Aσιά*, first attributed to Herodotus (about 440 BC) in reference to Anatolia or — in describing the Persian Wars — to the Persian Empire, in contrast to Greece and Egypt.

Herodotus comments that he is puzzled as to why three women's names are used to describe one enormous and substantial land mass (Europa, Asia, and Libya, referring to Africa), stating that most Greeks assumed that Asia was named after the wife of Prometheus (i.e. Hesione), but that the Lydians say it was named after Asias, son of Cotys, who passed the name on to a tribe in Sardis. Even before Herodotus.

Usage of the term soon became common in ancient Greece, and subsequently by the ancient Romans. Ancient and medieval European maps depict the Asian continent as a "huge amorphous blob" extending eastward. It was presumed in antiquity to end with India — the Greek king Alexander the Great believing he would reach the "end of the world" upon his arrival in the East.

Other Alternatives

Alternatively, the etymology of the term may be from the Akkadian word *(w) aSû(m)*, which means 'to go outside' or 'to ascend', referring to the direction of the sun at sunrise in the Middle East and also likely connected with the Phoenician word *asa* meaning east. This may be contrasted to a similar etymology proposed for *Europe*, as being from Akkadian *erebu(m)* 'to enter' or 'set' (of the sun).

T.R. Reid supports this alternative etymology, noting that the ancient Greek name must have derived from *asu*, meaning 'east' in Assyrian (*ereb* for *Europe* meaning 'west'). The ideas of *Occidental* (form Latin *Occidens* 'setting') and *Oriental* (from Latin *Oriens* for 'rising') are also European invention, synonymous with *Western* and *Eastern*. Reid further emphasizes that it explains the Western point of view of placing all the peoples and cultures of Asia into a single classification, almost as if there were a need for setting the distinction between Western and Eastern civilizations on the Eurasian continent. Ogura Kazuo and Tenshin Okakura are two Japanese outspoken figures over the subject.

However, this etymology is considered doubtful, because it does not explain how the term "Asia" first came to be associated with Anatolia, which is *west* of the Semitic-speaking areas, unless they refer to the viewpoint of a Phoenician sailor sailing through the straits between the Mediterranean Sea and the Black Sea.

Definition and Boundaries

Physical Geography

Medieval Europeans considered Asia as a continent a distinct landmass. The European concept of the three continents in the Old World goes back to Classical Antiquity, but during the Middle Ages was notably due to 7th century Spanish scholar Isidore of Sevilla. The demarcation between Asia and Africa (to the southwest) is the Isthmus of Suez and the Red Sea. The boundary between Asia and Europe is conventionally considered to run through the Dardanelles, the Sea of Marmara, the Bosporus, the Black Sea, the Caucasus Mountains, the Caspian Sea, the Ural River to its source and the Ural Mountains to the Kara Sea near Kara, Russia. While this interpretation of tripartite continents (i.e., of Asia, Europe and Africa) remains common in modernity, discovery of the extent of Africa and Asia have made this definition somewhat anachronistic. This is especially true in the case of Asia, which has several regions that would be considered distinct landmasses if these criteria were used (for example, Southern Asia and Eastern Asia).

In the far northeast of Asia, Siberia is separated from North America by the Bering Strait. Asia is bounded on the south by the Indian Ocean (specifically, from west to east, the Gulf of Aden, Arabian Sea and Bay of Bengal), on the east by the waters of the Pacific Ocean (including, counter clockwise, the South China Sea, East China Sea, Yellow Sea, Sea of Japan, Sea of Okhotsk and Bering Sea) and on the north by the Arctic Ocean. Australia (or Oceania) is to the southeast.

Some geographers do not consider Asia and Europe to be separate continents, as there is no logical physical separation between them. For example, Sir Barry Cunliffe, the emeritus professor of European archeology at Oxford, argues that Europe has been geographically and culturally merely "the western excrescence of the continent of Asia." Geographically, Asia is the major eastern constituent of the continent of Eurasia with Europe being a northwestern peninsula of the landmass – or of Afro-Eurasia: geologically, Asia, Europe and Africa comprise a single continuous landmass (save the Suez Canal) and share a common continental shelf. Almost all of Europe and most of Asia sit atop the Eurasian Plate, adjoined on the south by the Arabian and Indian Plate and with the easternmost part of Siberia (east of the Cherskiy Range) on the North American Plate.

In geography, there are two schools of thought. One school follows historical convention and treats Europe and Asia as different continents, categorizing subregions within them for more detailed analysis. The other school equates the word "continent" with a geographical region when referring to Europe, and use the term "region" to describe Asia in terms of physiography. Since, in linguistic terms, "continent" implies a distinct landmass, it is becoming increasingly common to substitute the term "region" for "continent" to avoid the problem of disambiguation altogether.

Given the scope and diversity of the landmass, it is sometimes not even clear exactly what "Asia" consists of. Some definitions exclude Turkey, the Middle East, Central Asia and Russia while only considering the Far East, Southeast Asia and the Indian subcontinent to compose Asia, especially in the United States after World War II. The term is sometimes used more narrowly in reference to the Asia-Pacific region, which does not include the Middle East, South Asia or Russia, but does include islands in the Pacific Ocean—a number of which may also be considered part of Australasia or Oceania, although Pacific Islanders are not considered Asian.

Economy

The economy of Asia comprises more than 4 billion people (60% of the world population) living in 46 different states. Six further states lie partly in Asia, but are considered to belong to another region economically and politically. As in all world regions, the wealth of Asia differs widely between, and within, states. This is due to its vast size, meaning a huge range of differing cultures, environments, historical ties and government systems. The largest economies in Asia in terms of nominal GDP are Japan, China, India, South Korea, Indonesia and Iran. Economies range from Japan, as the world's second largest economy by nominal GDP, to Cambodia as one of the poorest. In terms of GDP by purchasing power parity, China has the largest economy in Asia and the second largest economy in the world, followed by Japan and India as the world's third and fourth largest economies respectively.

Wealth (if measured by GDP per capita) is mostly concentrated in east Asian territories such as Hong Kong, Japan, South Korea, and Taiwan, as well in oil rich Middle Eastern countries such as Iran, Saudi Arabia, Qatar, United Arab Emirates. Asia, with the exception of Japan and South Korea, is currently undergoing rapid growth and industrialization spearheaded by China and India-the two fastest growing major economies in the world. While east Asian and southeast Asian countries generally rely on manufacturing and trade for growth, countries in the Middle East depend more on the production of commodities, principally oil, for economic growth. Over the years, with rapid economic growth and large trade surplus with the rest of the world, Asia has accumulated over US$4 trillion of foreign exchange reserves-more than half of the world's total.

Economic Development

Ancient and Medieval Times

Asia was rich in the ancient times. China and India alternated in being the largest economies in the world from about 1 A.D to 1800 A.D. China was a major economic power and attracted many to the east, and for many the legendary wealth and prosperity of the ancient culture of India personified Asia, attracting European commerce, exploration and colonialism. The accidental discovery of America by Columbus in search for India demonstrates this deep fascination. The Silk Road became the main East-West trading route in the Asian hitherland while the Straits of Malacca stood as a major sea route.

Pre-1945

Prior to World War II, most of Asia was under colonial rule. Only relatively few states managed to remain independent in the face of constant pressure exerted by European power. Such examples are Siam and Japan.

Japan in particular managed to develop its economy due to a reformation in the 19th century. The reformation was comprehensive and is today known as the Meiji Restoration. The Japanese economy continued to grow well into the 20th century and its economic growth created various shortages of resources essential to economic growth. As a result the Japanese expansion began with a great part of Korea and China annexed, thus allowing the Japanese to secure strategic resources.

At the same time, Southeast Asia was prospering due to trade and the introduction of various new technologies of that time. The volume of trade continued to increase with the opening of the Suez Canal in the 1860s. Manila had its gallion or Manila galleon wherein products from the Philippines were traded to Europe. The Philippines was the first Asian country to trade with Latin America via Acapulco. Tobacco, coconut, corn, and sugar trade was the most in demand during that time. Singapore, founded in 1819, rose to prominence as trade between the east and the west increased at an incredible rate. The British colony of Malaya, now part of Malaysia, was the world's largest producer of tin and rubber. The Dutch East Indies, now Indonesia, on the other hand, was known for its spices production. Both the British and the Dutch created their own trading companies to manage their trade flow in Asia. The British created the British East India Company while the Dutch formed Dutch East India Company. Both companies maintained trade monopolies of their respective colonies.

In 1908, crude oil was first discovered in Persia, modern day Iran. Afterwards, many oil fields were discovered and it was learnt later that the Mideast possesses the world's largest oil stocks. This made the rulers of the Arab nations very rich though the socioeconomic development in that region lagged behind.

In the early 1930s, the world underwent a global economic depression, today known as the Great Depression. Asia was not spared, and suffered the same pain as Europe and the United States. The volume of trade decreased dramatically all around Asia and indeed the world. With falling demand, prices of various goods starting to fall and further impoverished locals and foreigners alike. In 1941, Japan invaded Malaya and thus began World War II in Asia.

1945-1990

Following World War II, the People's Republic of China and India, which account for half of the population of Asia, adopted socialist policies to promote their domestic economy. These policies limited the economic growth of the region. In contrast, the economies of superiors Japan, South Korea and the other tigers Taiwan, Singapore, and Hong Kong—were economic successes, and the only successful economies outside of North America, Western Europe and Australia. The Philippines from the post-World War II until the late 1960s had the second largest economy in Asia. The Philippine economy during the 1980s was marked by stagnant growth as a result of dictatorship and martial law.

One of the most pronounced Asian economic phenomenons during this time-the Japanese post-war economic miracle greatly impacted the rest of the world. After World War II, under central guidance from the Japanese government, the entire economy was undergoing a remarkable restructuring. Close cooperation between the government, corporations and banks facilitated easy access to much-needed capital, and large conglomerates known as *keiretsu* spurred horizontal and vertical integration across all industries, keeping out foreign competition. These policies, in addition to an abandonment of military spending, worked phenomenally well. Japanese corporations as a result exported and still export massive amounts of high quality products from The Land of The Rising Sun.

Another amazing economic success story is that of South Korea's, also referred to as the Miracle on the Han River. The country was left impoverished after the Korean War, yet was able to recover at double digit percentiles. Many conglomerates, also known as Chaebols, such as Samsung, LG, Hyundai, Kia, SK, and more grew tremendously during this period. South Korea has now become the most wired country in the world.

Taiwan and Hong Kong experienced rapid growth up till the 1990s. Taiwan became, and still remains one of the main centers of consumer electronics R&D as well as manufacturing. However, unlike in Japan and South Korea, the bulk of Taiwan's economy is dependent on small to medium sized businesses. Hong Kong, on the other hand, experienced rapid growth in the financial sector due to liberal market policies, with many financial institutions setting up their Asian headquarters in Hong Kong. Till today, Hong Kong has been ranked as the world's freest economy for many years running, and it remains among one of the world's top 5 leading financial centers. This period was also marked by military conflict. Wars driven by the Cold War, notably in Vietnam and Afghanistan, wrecked the economies of these respective nations. When the Soviet Union collapsed in 1990-91, many Central Asian states were cut free and were forced to adapt to pressure for democratic and economic change. Also, several of the USSR's allies lost valuable aid and funding.

1991-2007

After the liberalization of the economy of India, undertaken by then finance minister and current Prime Minister of India, Dr. Manmohan Singh, the Indian economy coupled with the Chinese economy to power Asia into being one of the hotspots for world trade. The Chinese economy was already booming under the economic measures undertaken by Deng Xiaoping, in the 1980s, and continuing under Jiang Zemin in the 1990s. In 2007, China's economic growth rate exceeded 11% while India's growth rate increased to around 9%. One of the factors was the sheer size of the population in this region.

Meanwhile, Thailand, Malaysia and Indonesia emerged as the new Asian tigers with their GDPs growing well above 7% per year in the 1980s and the 90s. Their economies were mainly driven by growing exports. The Philippines only began to open up its stagnated economy in the early 1990s. Vietnam's economy began to grow in 1995, shortly after the United States and Vietnam restored economic and political ties. Throughout the 1990s, the manufacturing ability and cheap labour markets in Asian developing nations allowed companies to establish themselves in many of the industries previously dominated by companies from developed nations. Asia became one of the largest sources of automobiles, machinery, audio equipment and other electronics.

At the end of 1997, Thailand was hit by currency speculators, and the value of the Baht along with its annual growth rate fell dramatically. Soon after, the crisis spread to Indonesia, Malaysia, South Korea, Hong Kong, Singapore and many other Asian economies, resulting in great economic damage on the affected countries (Japan largely escaped the crisis). In fact, some of the economies, most notably those of Thailand, Indonesia, and South Korea actually contracted. This later would be known as the Asian financial crisis. By 1999, most countries had already recovered from the crisis.

In 2004, parts of Sumatra and South Asia were severely damaged by an earthquake and the subsequent tsunami. The natural disaster wiped out huge amounts of infrastructure throughout the affected area and displaced millions.

Future

Asia's large economic disparities are a source of major continuing tension in the region. While global economic powers Japan, China, India, and South Korea continue powering through, and Indonesia, The Philippines, Malaysia, Thailand, and Vietnam have entered the path to long-term growth, regions right next to these countries are in need of severe assistance.

Given the large number cheap and amply available labour in the region, particularly in China and India, where large workforces provide an economical advantage over other countries, the rising standard of living will eventually lead to a slow-down. Asia is also riddled with political problems that threaten not just the economies, but the general stability of the region and world. The nuclear neighbours—Pakistan and India—constantly pose a threat to each other, causing their governments to heavily invest in military spending.

Military intervention by the United States in Iraq and Afghanistan has also inflamed extremism and resulted in several terrorist attacks in a number of Asian countries. Another impending crisis is the depletion of oil reserves in the Middle East. Most of these economies have traditionally been over-dependent on oil and have had difficulty establishing another pillar in their economies. Yet another potential global danger posed by the economy of Asia is the growing accumulation of foreign exchange reserves. The countries/regions with the largest foreign reserves are mostly in Asia-China (Mainland-$988 billion & Hong Kong-$130 billion, September 2006), Japan ($881 billion, September 2006), Russia ($412 billion, June 2009), India ($277 billion, Mar 2010), Taiwan ($261 billion, September 2006), the Republic of Korea ($228 billion, September 2006), Singapore ($129 billion, June 2005). This increasingly means that the interchangeability of the Euro, USD, and GBP are heavily influenced by Asian central banks. Some economists in the western countries see this as a bad thing, prompting their respective governments to take action.

The economies of Asia are expected to be unequally divided for a long period of time. East Asian nations such as economic leaders Japan, China and South Korea will continue to flourish. Japanese products such as Sony and Hitachi are commanding premium prices in the western world. South Korean conglomerates Samsung and LG are respectively the second and fifth largest in Asia in terms of annual revenues. These two rivalling countries are expected to be joined by new economic competitors such as China and India. On the other hand, the Middle East and a few parts of South East Asia are will be in a state of trouble.

Economic Sectors

Primary Sector

Asia is by a considerable margin the largest continent in the world, and is rich in natural resources. The vast expanse of the former Soviet Union, particularly that of Russia, contains a huge variety of metals, such as gold, iron, lead, titanium, uranium, and zinc. These metals are mined, but inefficiently due to continued use of poorly maintained, obsolete machinery left over from the communist era. Nevertheless, profits are high due to a commodity price boom in 2003/2004 caused largely by increased demand in China. Oil is Southwest Asia's most important natural resource. Saudi Arabia, Iraq, and Kuwait are rich in oil reserves and have benefited from recent oil price escalations. Asia is home to some four billion people, and thus has a well established tradition in agriculture. High productivity in agriculture, especially of rice, allows high population density of many countries such as Bangladesh, Pakistan, southern China, Cambodia, India, and Vietnam. Agriculture constitutes a high portion of land usage in warm and humid areas of Asia. Many hillsides are farmed in a *terrace* method to boost arable land. The main agricultural products in Asia include rice and wheat. Opium is one of major cash crops in Central and Southeast Asia, particularly in Afghanistan, though its production is prohibited everywhere. Forestry is extensive throughout Asia except Southwest and Central Asia, with many of the items of furniture sold in the developed nations made out of Asian timber. Fishing is a major source of food, particularly in Japan.

Secondary Sector

The manufacturing sector in Asia has traditionally been strongest in the East region-particularly in China, Taiwan, Japan, South Korea and Singapore. The industry varies from manufacturing cheap low value goods such as toys to high-tech added value goods such as computers, CD players, Games consoles, mobile phones and cars. Major Asian manufacturing companies are mostly based in either South Korea or Japan. They include Samsung, Hyundai, LG, and Kia from South Korea, and Sony, Toyota, Toshiba, and Honda from Japan. Many developed-nation firms from Europe, North America, Japan and South Korea have significant operations in the developing Asia to take advantage of the abundant supply of cheap labour. One of the major employers in manufacturing in Asia is the textile industry. Much of the world's supply of clothing and footwear now originates in Southeast Asia and South Asia, particularly in Vietnam, China, India, Thailand, Bangladesh, Pakistan, and Indonesia.

Tertiary Sector

Asia has six important financial centers, located in Dubai, Mumbai, Hong Kong, Shanghai, Singapore and Tokyo. India has been one of the greatest beneficiaries of the economic boom. The country has emerged as one of the world's largest exporters of software and other information technology related services. World class Indian software giants such as Infosys, HCL, Wipro, Mahindra Satyam and TCS have emerged as the world's most sought after service providers. Call centers are also becoming major employers in India and Philippines due to the availability of many well educated English speakers. Here again India holds close to 60% of the trade share. The rise of the Business Process Outsourcing (BPO) industry has seen the rise of India and China as the other financial centers. Experts believe that the current centre of financial activity is moving toward "Chindia"-a name used for jointly referring to China and India-with Shanghai and Mumbai becoming major financial hubs in their own right. Other growing technological and financial hubs include Dhaka (Bangladesh), Bangalore (India), Chennai (India), New Delhi (India), Hyderabad (India), Jakarta (Indonesia), Kuala Lumpur (Malaysia), Kolkata (India), Karachi (Pakistan), Manila (Philippines) and Bangkok (Thailand).

China was the largest and most advanced economy on earth for much of recorded history, until the British Empire (excluding India) overtook it in the mid 19th century. Japan has had for only several decades after WW2 the largest economy in Asia and second-largest of any single nation in the world, after surpassing the Soviet Union (measured in net material product) in 1986 and Germany in 1968. (NB: A number of supernational economies are larger, such as the European Union (EU), the North American Free Trade Agreement (NAFTA) or APEC).

In the late 1980s and early 1990s, Japan's GDP was almost as large (current exchange rate method) as that of the rest of Asia combined. In 1995, Japan's economy nearly equaled that of the USA to tie as the largest economy in the world for a day, after the Japanese currency reached a record high of 79 yen/dollar. Economic growth in Asia since World War II to the 1990s had been concentrated in Japan as well as the four regions of South Korea, Taiwan, Hong Kong and Singapore located in the Pacific Rim, known as the Asian tigers, which have now all received developed country status, having the highest GDP per capita in Asia.

It is forecasted that India will overtake Japan in terms of Nominal GDP by 2020. In terms of GDP per capita, both nominal and PPP-adjusted, South Korea will become the second wealthiest country in Asia by 2025, overtaking Germany, the United Kingdom and France. By 2050, according to a 2006 report by Price Waterhouse Cooper, China will have the largest economy in the world (43% greater than the United States when PPP adjusted, although perhaps smaller than the United States in nominal terms).

Trade blocs:

- Asia-Pacific Economic Cooperation
- Asia-Europe Economic Meeting
- Association of Southeast Asian Nations
- Gulf Cooperation Council
- Closer Economic Partnership Arrangement
- Commonwealth of Independent States
- South Asian Association for Regional Cooperation.

Natural Resources

Asia is the largest continent in the world by a considerable margin, and it is rich in natural resources, such as petroleum, forests, fish, water, rice, copper and silver.

Manufacturing

Manufacturing in Asia has traditionally been strongest in East and Southeast Asia, particularly in mainland China, Taiwan, South Korea, Japan, India, Philippines and Singapore. Japan and South Korea continue to dominate in the area of multinational corporations, but increasingly mainland China, and India are making significant inroads. Many companies from Europe, North America, South Korea and Japan have operations in Asia's developing countries to take advantage of its abundant supply of cheap labour and relatively developed infrastructure.

Financial and other Services

Asia has four main financial centres: Tokyo, Hong Kong, Singapore and Shanghai. Call centres and business process outsourcing (BPOs) are becoming major employers in India and the Philippines due to the availability of a large pool of highly skilled, English-speaking workers. The increased use of outsourcing has assisted the rise of India and the China as financial centres. Due to its large and extremely competitive information technology industry, India has become a major hub for outsourcing.

Early History

The coastal periphery was home to some of the world's earliest known civilizations, each of them developing around fertile river valleys. The civilizations in Mesopotamia, the Indus Valley and the Huanghe shared many similarities. These civilizations may well have exchanged technologies and ideas such as mathematics and the wheel. Other innovations, such as writing, seem to have been developed individually in each area. Cities, states and empires developed in these lowlands.

The central steppe region had long been inhabited by horse-mounted nomads who could reach all areas of Asia from the steppes. The earliest postulated expansion out of the steppe is that of the Indo-Europeans, who spread their languages into the Middle East, South Asia, and the borders of China, where the Tocharians resided. The northernmost part of Asia, including much of Siberia, was largely inaccessible to the steppe nomads, owing to the dense forests, climate and tundra. These areas remained very sparsely populated.

The centre and the peripheries were mostly kept separated by mountains and deserts. The Caucasus and Himalaya mountains and the Karakum and Gobi deserts formed barriers that the steppe horsemen could cross only with difficulty. While the urban city dwellers were more advanced technologically and socially, in many cases they could do little in a military aspect to defend against the mounted hordes of the steppe. However, the lowlands did not have enough open grasslands to support a large horsebound force; for this and other reasons, the nomads who conquered states in China, India, and the Middle East often found themselves adapting to the local, more affluent societies.

Languages and Literature

Asia is home to several language families and many language isolates. Most Asian countries have more than one language that is natively spoken. For instance, according to Ethnologue, more than 600 languages are spoken in Indonesia, more than 800 languages spoken in India, and more than 100 are spoken in the Philippines. China has many languages and dialects in different provinces.

Nobel Prizes

The polymath Rabindranath Tagore, a Bengali poet, dramatist, and writer from Santiniketan, now in West Bengal, India, became in 1913 the first Asian Nobel laureate. He won his Nobel Prize in Literature for notable impact his prose works and poetic thought had on English, French, and other national literatures of Europe and the Americas. He is also the writer of the national anthems of Bangladesh and India.

Tagore is said to have named another Bengali Indian Nobel prize winner, the 1998 laureate in Economics, Amartya Sen. Sen's work has centered around global issues including famine, welfare, and third-world development. Amartya Sen was Master of Trinity College, Cambridge University, UK, from 1998–2004, becoming the first Asian to head an 'Oxbridge' College.

Other Asian writers who won Nobel Prizes include Yasunari Kawabata (Japan, 1966), Kenzaburô (Japan, 1994), Gao Xingjian (People's Republic of China, 2000) and Orhan Pamuk (Turkey, 2006).

Also, Mother Teresa of India and Shirin Ebadi of Iran were awarded the Nobel Peace Prize for their significant and pioneering efforts for democracy and human rights, especially for the rights of women and children. Ebadi is the first Iranian and the first Muslim woman to receive the prize. Another Nobel Peace Prize winner is Aung San Suu Kyi from Burma for her peaceful and non-violent struggle under a military dictatorship in Burma. She is a nonviolent pro-democracy activist and leader of the National League for Democracy in

Burma(Myanmar) and a noted prisoner of conscience. She is a Buddhist and was awarded the Nobel Peace Prize in 1991.

Sir C.V.Raman is the first Asian to get a Nobel prize in Sciences. He won the Nobel Prize in Physics "for his work on the scattering of light and for the discovery of the effect named after him".

Other Asian Nobel Prize winners include Subrahmanyan Chandrasekhar, Abdus Salam, Shmuel Yosef Agnon, Robert Aumann, Menachem Begin, Aaron Ciechanover, Avram Hershko, Daniel Kahneman, Shimon Peres, Yitzhak Rabin, Yaser Arafat, Jose Ramos Horta and Bishop Carlos Filipe Ximenes Belo of Timor Leste, Kim Dae-jung, and thirteen Japanese scientists. Most of the said awardees are from Japan and Israel except for Chandrasekhar and Raman (India), Salam (Pakistan), Arafat (Palestinian Territories) and Kim (South Korea).

In 2006, Dr. Muhammad Yunus of Bangladesh was awarded the Nobel Peace Prize for the establishment of Grameen Bank, a community development bank that lends money to poor people, especially women in Bangladesh. Dr. Yunus received his Ph.D. in economics from Vanderbilt University, United States. He is internationally known for the concept of micro credit which allows poor and destitutes with little or no collateral to borrow money. The borrowers typically pay back money within the specified period and the incidence of default is very low.

The Dalai Lama has received approximately eighty-four awards over his spiritual and political career. On 22 June 2006, he became one of only four people ever to be recognized with Honorary Citizenship by the Governor General of Canada. On 28 May 2005, he received the Christmas Humphreys Award from the Buddhist Society in the United Kingdom. Most notable was the Nobel Peace Prize, presented in Oslo, Norway on 10 December 1989.

Mythology

Asian mythology is complex and diverse. The story of the Great Flood for example, as presented to Christians in the Old Testament, is first found in Mesopotamian mythology, in the *Epic of Gilgamesh*. Hindu mythology tells about an avatar of the God Vishnu in the form of a fish who warned Manu of a terrible flood. In ancient Chinese mythology, Shan Hai Jing, the Chinese ruler Da Yu, had to spend 10 years to control a deluge which swept out most of ancient China and was aided by the goddess Nuwa who literally fixed the broken sky through which huge rains were pouring.

Religions

Almost all Asian religions have philosophical character and Asian philosophical traditions cover a large spectrum of philosophical thoughts and writings. Indian philosophy includes Hindu philosophy and Buddhist philosophy. They include elements of non-material pursuits, whereas another school of thought from India, Carvaka, preached the enjoyment of material world. Christianity is also present in most Asian countries.

Abrahamic

The Abrahamic religions of Judaism, Christianity, Islam and Bahai Faith originated in West Asia. Judaism, the oldest of the Abrahamic faiths, is practiced primarily in Israel (which has the world's largest Jewish population), though small communities exist in other countries, such as the Bene Israel in India. In the Philippines and East Timor, Roman Catholicism is the predominant religion; it was introduced by the Spaniards and the Portuguese, respectively. In Armenia, Cyprus, Georgia and Russia, Eastern Orthodoxy is the predominant religion. Various Christian denominations have adherents in portions of the Middle East, as well as China and India.

The world's largest Muslim community (within the bounds of one nation) is in Indonesia. South Asia (mainly Pakistan, India and Bangladesh) holds 30% of Muslims. There are also significant Muslim populations in China, Iran, Malaysia, southern Philippines (Mindanao), Russia and most of West Asia and Central Asia. The Bahai Faith originated in Asia, in Iran (Persia), and spread from there to the Ottoman Empire, Central Asia, India, and Burma during the lifetime of Bahaullah. Since the middle of the 20th Century, growth has particularly occurred in other Asian countries, because the Bahai Faith's activities in many Muslim countries has been severely suppressed by authorities.

Dharmic and Taoist

The religions of Hinduism, Buddhism, Jainism and Sikhism originated in India, South Asia. In East Asia, particularly in China and Japan, Confucianism, Taoism and Zen Buddhism took shape.

Asia and its Geography

Asia is the central and eastern part of Eurasia, comprising approximately fifty countries. It has an area, including islands, of roughly 49,694,700 km^2. Asia is joined to Africa by the Isthmus of Suez and to Europe by a long border generally following the Ural Mountains.

Medieval Europeans considered Asia as a continent, a distinct landmass. The European concept of the three continents in the Old World goes back to classical antiquity with the etymology of the word rooted in the ancient Near and Middle East. The demarcation between Asia and Africa is the isthmus of Suez and the Red Sea.

The boundary between Asia and Europe is commonly believed to run through the Dardanelles, the Sea of Marmara, the Bosporus, the Black Sea, the Caucasus Mountains, the Caspian Sea, the Ural River to its source, and the Ural Mountains to the Kara Sea near Kara, Russia. However, modern discovery of the extent of Africa and Asia made this definition rather anachronistic, especially in the case of Asia, which would have several regions that would be considered distinct landmasses if these criteria were used (for example, South Asia and East Asia).

Geologists and physical geographers no longer consider Europe and Asia to be separate continents. It is either defined in terms of geological landmasses (physical geography) or tectonic plates (geology). In the former case, Europe is a western peninsula of Eurasia or the Africa-Eurasia landmass. In the latter, Europe and Asia are parts of the Eurasian plate, which excludes the Arabian and Indian tectonic plates.

In human geography, there are two schools of thought. One school follows historical convention and treats Europe and Asia as different continents, categorizing Europe, East Asia (the Orient), South Asia (British India), and the Middle East (Arabia and Persia) as specific regions for more detailed analysis. The other schools equate the word "continent" in terms of geographical region when referring to Europe, and use the term "region" to describe Asia in terms of physical geography. It is becoming increasingly common to substitute the term "region" for "continent" because in linguistic terms, "continent" implies a distinct landmass.

There is much confusion in European languages with the term "Asian", which almost always refers to a subcategory of people from Asia rather than referring to "Asian" defined in term of "Asia", because a category implies homogeneity. In American English, Asian refers to East Asians, while in British English, Asian refers to South Asians. Some definitions of Asia exclude Turkey, the Middle East, or Russia. The term is sometimes used more strictly in reference to Asia Pacific, which does not include the Middle East or Russia, but does include islands in the Pacific Ocean — many of which are considered part of Australasia or Oceania. Asia contains the Indian subcontinent, Arabian peninsula, as well as a piece of the North American plate in Siberia.

Main Geographical Features

The mean elevation of the continent is 950 m (3,117 ft.), the highest of any in the world. The plateau and mountainous areas broadly sweep SW-NW across Asia, climaxing in the high Tibetan Plateau, rising to the highest peaks in the world in the Himalayas. To the northwest lie plains, while to the south lie the geologically distinct areas of the Arabian Peninsula, Indian subcontinent and Malay Peninsula. Large numbers of islands lie southeast of the continent.

Besides its mainland, Asia includes a large number of islands, including some of the world's largest islands, such as Borneo and Sumatra, and some of the world's most populous islands, such as Java and Honshu. Other prominent islands include Bali, Madura and Sulawesi of Indonesia; Hokkaido, Shikoku, Kyushu and Okinawa of Japan; the Andaman and Nicobar of India; Luzon and Mindanao of the Philippines; Ko Pha Ngan and Ko Samui of Thailand; as well as Sri Lanka, Maldives, Singapore, Hong Kong Island and Sakhalin.

The nations of Indonesia, Brunei, East Timor, Singapore, Japan, Philippines, Taiwan, Sri Lanka, Maldives and Cyprus are solely made up of one or more islands, and have no territory on the mainland.

Extreme Points

The following lists the points of Asia that are furthest north, south, east and west. Some of these points are open to debate, as the definition of Asia is varied.

Including Islands

- Northernmost point — Arctic Cape, Komsomolets Island, Severnaya Zemlya, Russia. (81°13' N)
- Southernmost point — Pamana Island, a small island off Rote Island, Indonesia, has been described as Asia's southernmost point.. However, if the Cocos (Keeling) Islands are included as part of Asia, then South Island (12°04'S) is the southernmost point.
- Westernmost point — Cape Baba, Turkey (26°4'E). This is the westernmost point of the Asian part of Turkey. Although Turkey governs some more westerly islands in the Aegean and Mediterranean Seas, they are considered part of Europe.
- Easternmost point — Big Diomede, Russia (169°0'W) The International Date Line runs between the Russian Big Diomede and the neighbouring U.S.-governed Little Diomede.
- Pole of inaccessibility : Dzoosotoyn Elisen Desert, Xinjiang, China.

Mainland

- Northernmost point — Cape Chelyuskin, Russia (77°43'N)
- Southernmost point — Cape Piai, Malaysia (1°16'N)
- Westernmost point — Cape Baba, Turkey (26°4'E
- Easternmost point — Cape Dezhnev (East Cape), Russia (169°40'W).

Geographical Regions

Asia is a subregion of Eurasia. For further subdivisions based on that term, see North Eurasia and Central Eurasia.

- Central Asia
- Iranian Plateau
- East Asia
- Far East
- North Asia
- South Asia (also Indian subcontinent)
- Southeast Asia
- Southwest Asia (or West Asia).

Central Asia

There is no absolute consensus in the usage of this term. Usually, Central Asia includes:

- The Central Asian republics of Kazakhstan (excluding its small European territory), Uzbekistan, Tajikistan, Turkmenistan and Kyrgyzstan
- Afghanistan, Mongolia, Iran and the western regions of China are also sometimes included
- Former Soviet states in the Caucasus region.

Central Asia is currently geopolitically important because international disputes and conflicts over oil pipelines, Nagorno-Karabakh, and Chechnya, as well as the presence of U.S. and U.K. military forces in Afghanistan.

East Asia

This area includes:

- Japan
- North and South Korea on the Korean Peninsula.
- China, but sometimes only the eastern regions

- Taiwan
- Mongolia.

More informally, Southeast Asia is included in East Asia on some occasions.

North Asia

This term is rarely used by geographers, but usually it refers to the larger Asian part of Russia, also known as Siberia. Sometimes the northern parts of other Asian nations, such as Kazakhstan or Mongolia, are also included in North Asia. The term "Northern Eurasia" is sometimes used as an apolitical term that includes all the nations of the former U.S.S.R.

Southeast Asia

This region contains the Malay Peninsula, Indochina and islands in the Indian Ocean and Pacific Ocean. The countries it contains are:

- In mainland Southeast Asia, the countries Burma, Thailand, Laos, Cambodia and Vietnam.
- In maritime Southeast Asia, the countries of Malaysia, Brunei, the Philippines, Singapore and Indonesia (some of the Indonesian islands also lie in the Melanesia region of Oceania). East Timor (also Melanesian) is sometimes included too.

Malaysia is divided in two by the South China Sea, and thus has both a mainland and an island part.

West Asia

West Asia is often called the *Middle East* by Europeans and Americans. *Middle East* (to some interpretations) is often used to also refer to some countries in North Africa. West Asia can be further divided into:

- Anatolia (i.e. Asia Minor), constituting the Asian part of Turkey
- The island nation of Cyprus in the Mediterranean Sea
- The Levant or Near East, which includes Syria, Israel, Jordan, Lebanon, Iraq and the Asian portion of Egypt
- The Arabian Peninsula, including Saudi Arabia, United Arab Emirates, Bahrain, Qatar, Oman, Yemen and Kuwait
- The Caucasus region (which straddles both Asia and Europe), namely Transcaucasia, including a small portion of Russia and, arguably, most if not all of Georgia, Armenia, and Azerbaijan
- The Iranian Plateau, including Iran and Afghanistan.

India and Southeastern Asia

India, officially the Republic of India is a country in South Asia. It is the seventh-largest country by geographical area, the second-most populous country with 1.18 billion people, and the most populous democracy in the world. Mainland India is bounded by the Indian Ocean on the south, the Arabian Sea on the west, and the Bay of Bengal on the east; and it is bordered by Pakistan to the west; China, Nepal, and Bhutan to the north; and Bangladesh and Burma to the east. India is in the vicinity of Sri Lanka, and the Maldives in the Indian Ocean, and its Andaman and Nicobar Islands are also in the vicinity of the Indonesian island of Sumatra in the Andaman Sea. India has a coastline of 7,517 kilometres (4,700 mi).

Home to the Indus Valley Civilisation and a region of historic trade routes and vast empires, the Indian subcontinent was identified with its commercial and cultural wealth for much of its long history. Four major religions, Hinduism, Buddhism, Jainism and Sikhism originated here, while Zoroastrianism, Judaism, Christianity and Islam arrived in the first millennium CE and shaped the region's diverse culture. Gradually annexed by the British East India Company from the early eighteenth century and colonised by the United Kingdom from the mid-nineteenth century, India became an independent nation in 1947 after a struggle for independence that was marked by widespread non-violent resistance.

India is a federal constitutional republic consisting of 28 states and seven union territories with a parliamentary system of democracy. The Indian economy is the world's eleventh largest economy by nominal GDP and the fourth largest by purchasing power parity. Economic reforms since 1991 have transformed it into one of the fastest growing economies in the world. India is a nuclear weapons state and has the third-largest standing army in the world while its military expenditure ranks tenth.

India is considered to be a potential superpower, having a rapidly growing economy and growing political clout. India is one of main troop-contributing countries to United Nations peacekeeping operations. It is a member state of the United Nations and members of Commonwealth of Nations, the South Asian Association for Regional Cooperation, G15, G20, Asia Cooperation Dialogue, the Colombo Plan, Indian Ocean Rim Association for Regional Cooperation and the Non-aligned movement, and is one of the BRIC Countries. A pluralistic, multilingual and multiethnic society, India is also home to a diversity of wildlife in a variety of protected habitats.

Etymology

The name *India* is derived from *Indus*, which is derived from the Old Persian word *Hindu*, from Sanskrit, *Sindhu*, the historic local appellation for the Indus River. The ancient Greeks referred to the Indians as *Indoi*, the people of the Indus. The Constitution of India and common usage in various Indian languages also recognise *Bharat* as an official name of equal status. The name Bharat is derived from the name of the legendary king Bharata in Hindu Mythology. *Hindustan*, originally a Persian word for "Land of the Hindus" referring to northern India, is also occasionally used as a synonym for all of India.

History

Stone Age rock shelters with paintings at the Bhimbetka rock shelters in Madhya Pradesh are the earliest known traces of human life in India. The first known permanent settlements appeared over 9,000 years ago and gradually developed into the Indus Valley Civilisation, dating back to 3400 BCE in western India. It was followed by the Vedic period, which laid the foundations of Hinduism and other cultural aspects of early Indian society, and ended in the 500s BCE. From around 550 BCE, many independent kingdoms and republics known as the Mahajanapadas were established across the country.

In the third century BCE, most of South Asia was united into the Maurya Empire by Chandragupta Maurya and flourished under Ashoka the Great. From the third century CE, the Gupta dynasty oversaw the period referred to as ancient "India's Golden Age". Empires in Southern India included those of the Chalukyas, the Cholas and the Vijayanagara Empire. Science, technology, engineering, art, logic, language, literature, mathematics, astronomy, religion and philosophy flourished under the patronage of these kings.

Following invasions from Central Asia between the 10th and 12th centuries, much of North India came under the rule of the Delhi Sultanate and later the Mughal Empire. Under the rule of Akbar the Great, India enjoyed much cultural and economic progress as well as religious harmony. Mughal emperors gradually expanded their empires to cover large parts of the subcontinent. However, in North-Eastern India, the dominant power was the Ahom kingdom of Assam, among the few kingdoms to have resisted Mughal subjugation. The first major threat to Mughal imperial power came from a Hindu Rajput king Maha Rana Pratap of Mewar in the 16th century and later from a Hindu state known as the Maratha confederacy, that ruled much of India in the mid-18th century.

From the 16th century, European powers such as Portugal, the Netherlands, France, and Great Britain established trading posts and later took advantage of internal conflicts to establish colonies in the country. By 1856, most of India was under the control of the British East India Company. A year later, a nationwide insurrection of rebelling military units and kingdoms, known as India's First War of Independence or the Sepoy Mutiny, seriously challenged the Company's control but eventually failed. As a result of the instability, India was brought under the direct rule of the British Crown.

In the 20th century, a nationwide struggle for independence was launched by the Indian National Congress and other political organisations. Indian leader Mahatma Gandhi led millions of people in several national campaigns of non-violent civil disobedience.

On 15 August 1947, India gained independence from British rule, but at the same time the Muslim-majority areas were partitioned to form a separate state of Pakistan. On 26 January 1950, India became a republic and a new constitution came into effect. Since independence, India has faced challenges from religious violence, casteism, naxalism, terrorism and regional separatist insurgencies, especially in Jammu and Kashmir and Northeast India. Since the 1990s terrorist attacks have affected many Indian cities. India has unresolved territorial disputes with the People's Republic of China, which, in 1962, escalated into the Sino-Indian War, and with Pakistan, which resulted in wars in 1947, 1965, 1971 and 1999. India is a founding member of the United Nations (as British India) and the Non-Aligned Movement. In 1974, India conducted an underground nuclear test and five more tests in 1998, making India a nuclear state. Beginning in 1991, significant economic reforms have transformed India into one of the fastest-growing economies in the world, increasing its global clout.

Government

India is federation with a parliamentary form of government, governed under the Constitution of India. It is a constitutional republic and representative democracy, "in which majority rule is tempered by minority rights protected by law." Federalism in India defines the power distribution between the centre and the states. The government is regulated by a checks and balances defined by Indian Constitution, which serves as the country's supreme legal document.

Constitution

The Constitution of India, the longest and the most exhaustive among constitutions of independent nations in the world, came into force on 26 January 1950. The preamble of the constitution defines India as a sovereign, socialist, secular, democratic republic. India has a bicameral parliament operating under a Westminster-style parliamentary system. Its form of government was traditionally described as being 'quasi-federal' with a strong centre and weaker states, but it has grown increasingly federal since the late 1990s as a result of political, economic and social changes.

President and Prime Minister

The President of India is the head of state elected indirectly by an electoral college for a five-year term. The Prime Minister is the head of government and exercises most executive power. Appointed by the President, the Prime Minister is by convention supported by the party or political alliance holding the majority of seats in the lower house of Parliament. The executive branch consists of the President, Vice-President, and the Council of Ministers (the Cabinet being its executive committee) headed by the Prime Minister. Any minister holding a portfolio must be a member of either house of parliament. In the Indian parliamentary system, the executive is subordinate to the legislature, with the Prime Minister and his Council being directly responsible to the lower house of the Parliament.

Legislature

The Legislature of India is the bicameral Parliament, which consists of the upper house called the Rajya Sabha (Council of States) and the lower house called the Lok Sabha (House of People). The Rajya Sabha, a permanent body, has 245 members serving staggered six year terms. Most are elected indirectly by the state and territorial legislatures in proportion to the state's population. 543 of the Lok Sabha's 545 members are directly elected by popular vote to represent individual constituencies for five year terms. The other two members are nominated by the President from the Anglo-Indian community if the President is of the opinion that the community is not adequately represented.

Judiciary

India has a unitary three-tier judiciary, consisting of the Supreme Court, headed by the Chief Justice of India, 21 High Courts, and a large number of trial courts. The Supreme Court has original jurisdiction over cases involving fundamental rights and over disputes between states and the Centre, and appellate jurisdiction over the High Courts. It is judicially independent, and has the power to declare the law and to strike down Union or State laws which contravene the

Constitution. The role as the ultimate interpreter of the Constitution is one of the most important functions of the Supreme Court.

Politics

India is the most populous democracy in the world. It has operated under a multi-party system for most of its history. For most of the years since independence, the federal government has been led by the Indian National Congress (INC). Politics in the states have been dominated by national parties like the INC, the Bharatiya Janata Party (BJP) and various regional parties. From 1950 to 1990, barring two brief periods, the INC enjoyed a parliamentary majority. The INC was out of power between 1977 and 1980, when the Janata Party won the election owing to public discontent with the state of emergency declared by the then Prime Minister Indira Gandhi. In 1989, a Janata Dal-led National Front coalition in alliance with the Left Front coalition won the elections but managed to stay in power for only two years. As the 1991 elections gave no political party a majority, the INC formed a minority government under Prime Minister P.V. Narasimha Rao and was able to complete its five-year term.

The years 1996–1998 were a period of turmoil in the federal government with several short-lived alliances holding sway. The BJP formed a government briefly in 1996, followed by the United Front coalition that excluded both the BJP and the INC. In 1998, the BJP formed the National Democratic Alliance (NDA) with several other parties and became the first non-Congress government to complete a full five-year term.

In the 2004 Indian elections, the INC won the largest number of Lok Sabha seats and formed a government with a coalition called the United Progressive Alliance (UPA), supported by various Left-leaning parties and members opposed to the BJP. The UPA again came into power in the 2009 general election; however, the representation of the Left leaning parties within the coalition has significantly reduced. Manmohan Singh became the first prime minister since Jawaharlal Nehru in 1962 to be re-elected after completing a full five-year term.

Foreign Relations and Military

Since its independence in 1947, India has maintained cordial relationships with most nations. It took a leading role in the 1950s by advocating the independence of European colonies in Africa and Asia. India is a member of the Commonwealth of Nations and a founding member of the Non-Aligned Movement. India was involved in two brief military interventions in neighbouring countries – Indian Peace Keeping Force in Sri Lanka and Operation Cactus in Maldives. After the Sino-Indian War and the Indo-Pakistani War of 1965, India's relationship with the Soviet Union warmed and continued to remain so until the end of the Cold War. India has fought two wars with Pakistan over the Kashmir dispute. A third war between India and Pakistan in 1971 resulted in the creation of Bangladesh (then East Pakistan). Additional skirmishes have taken place between the two nations over the Siachen Glacier. In 1999, India and Pakistan fought an undeclared war over Kargil.

In recent years, India has played an influential role in the SAARC and the WTO. India has provided as many as 55,000 Indian military and police personnel to serve in thirty-five UN peacekeeping operations across four continents. India is also an active participant in various mutlilateral forums, particularly the East Asia Summit and the G8+5. Recent overtures by the Indian government have strengthened relations with the United States and China. In the economic sphere, India has close relationships with other developing nations in South America, Asia and Africa.

India maintains the third-largest military force in the world, which consists of the Indian Army, Navy, Air Force and auxiliary forces such as the Paramilitary Forces, the Coast Guard, and the Strategic Forces Command. The official Indian defence budget for 2010 stood at US$31.9 billion (or 2.12% of GDP). According to a 2008 SIPRI report, India's annual military expenditure in terms of PPP stood at US$72.7 billion. The President of India is the supreme commander of the Indian Armed Forces. India maintains close defence cooperation with Russia, Israel and France, who are the chief suppliers of arms. Defence contractors, such as the Defence Research and Development Organisation (DRDO) and Hindustan Aeronautics (HAL), oversee indigenous development of sophisticated arms and military equipment, including ballistic missiles, fighter aircraft and main battle tanks, to reduce India's dependence on foreign imports.

India became a nuclear power in 1974 after conducting an initial nuclear test, known as the Operation Smiling Buddha, and carried out further underground testing in 1998. Despite criticism and military sanctions, India has consistently refused to sign the CTBT and the NPT. India maintains a "no first use" nuclear policy and is developing nuclear triad capability as a part of its "minimum credible deterrence" doctrine. On 10 October 2008, a civilian nuclear agreement between India and the United States was signed, prior to which India received waivers from the IAEA and the NSG which ended restrictions on nuclear technology commerce and recognized India as the world's *de facto* sixth nuclear weapons state.. On 12 March 2010, Russia signed with India a nuclear reactor deal which will build 16 nuclear reactors in India as part of defence and energy deals. On 28 June 2010, Canada signs with India a nuclear co-operation deal to promote and develop co-operation in civilian nuclear energy.

Geography of India

The geography of India describes the physical features of India, a country in South Asia that lies entirely on the Indian Plate in the northern portion of the Indo-Australian Plate. The country lies to the north of the equator between 8°4' and 37°6' north latitude and 68°7' and 97°25' east longitude. It is the seventh-largest country in the world, with a total land area of 3,287,263 square kilometres (1,269,219 sq mi). India measures 3,214 km (1,997 mi) from north to south and 2,993 km (1,860 mi) from east to west. It has a land frontier of 15,200 km (9,445 mi) and a coastline of 7,517 km (4,671 mi).

India is bounded to the southwest by the Arabian Sea, to the southeast by the Bay of Bengal and the Indian Ocean to the south. Kanyakumari constitutes the southern tip of the Indian peninsula, which narrows before ending in the Indian Ocean. The southernmost part of India is Indira Point in the Andaman and Nicobar Islands. The Maldives, Sri Lanka and Indonesia are island nations to the south of India with Sri Lanka separated from India by a narrow channel of sea formed by Palk Strait and the Gulf of Mannar. The territorial waters of India extend into the sea to a distance of 12 nautical miles (13.8 mi; 22.2 km) measured from the appropriate baseline.

The northern frontiers of India are defined largely by the Himalayan mountain range where its political boundaries with China, Bhutan, and Nepal lie. Its western borders with Pakistan lie in the Punjab Plain and the Thar desert. In the far northeast, the Chin Hills and Kachin Hills, deeply forested mountainous regions, separate India from Burma while its political border with Bangladesh is defined by the watershed region of the Indo-Gangetic Plain, the Khasi hills and Mizo Hills.

The Ganges is the longest river originating in India and forms the Indo-Gangetic Plain. The Ganges-Brahmaputra system occupies most of northern, central and eastern India, while the Deccan Plateau occupies most of southern India. Along its western frontier is the Thar Desert, which is the seventh-largest desert in the world.

Officially, India's highest point is K2 at 8,611 m (28,251 ft), though it lies in Gilgit-Baltistan, part of the disputed Kashmir region. Kanchenjunga in Sikkim at 8,598 m (28,209 ft) is the highest point within India's current geographic boundaries. Climate across India ranges from equatorial in the far south, to Alpine in the upper reaches of the Himalayas.

Geological Development

India is entirely contained on the Indian Plate, a major tectonic plate that was formed when it split off from the ancient continent Gondwanaland. About 90 million years ago, during the late Cretaceous Period, the Indian Plate began moving north at about 15 cm/yr (6 in/yr). About 50 to 55 million years ago, in the Eocene epoch of the Cenozoic Era, the plate collided with Asia after covering a distance of 2,000 to 3,000 km (1,243 to 1,864 mi), having moved faster than any other known plate.

In 2007, German geologists determined that the reason the India Plate moved so quickly is that it is only half as thick as the other plates which formerly constituted Gondwanaland. The collision with the Eurasian Plate along the modern border between India and Nepal formed the orogenic belt that created the Tibetan Plateau and the Himalayas. As of 2009, The India Plate is moving northeast at 5 cm/yr (2 in/yr), while the Eurasian Plate is moving north at only 2 cm/yr (0.8 in/yr). India is thus referred to as the "fastest continent." This is causing the Eurasian Plate to deform, and the India Plate to compress at a rate of 4 mm/yr (0.15 in/yr).

Political Geography

India is divided into twenty-eight states (further subdivided into districts) and seven union territories.

States:

1. Andhra Pradesh
2. Arunachal Pradesh
3. Assam
4. Bihar
5. Chhattisgarh
6. Goa
7. Gujarat
8. Haryana
9. Himachal Pradesh
10. Jammu and Kashmir

11. Jharkhand
12. Karnataka
13. Kerala
14. Madhya Pradesh
15. Maharashtra
16. Manipur
17. Meghalaya
18. Mizoram
19. Nagaland
20. Orissa
21. Punjab
22. Rajasthan
23. Sikkim
24. Tamil Nadu
25. Tripura
26. Uttar Pradesh
27. Uttarakhand
28. West Bengal.

Union Territories

A. Andaman and Nicobar Islands
B. Chandigarh
C. Dadra and Nagar Haveli
D. Daman and Diu
E. Lakshadweep
F. National Capital Territory of Delhi
G. Puducherry.

India's borders run a total length of 15,106.70 km (9,387 mi). Its borders with Pakistan and Bangladesh were delineated according to the Radcliffe Line, which was created in 1947 during Partition of India. Its western border with Pakistan extends up to 3,323 km (2,065 mi), dividing the Punjab region and running along the boundaries of the Thar Desert and the Rann of Kutch. Both nations delineated a Line of Control (LoC) to serve as the informal boundary between the Indian and Pakistan-administered areas of Kashmir. According to India's claim, it shares a 106 km (66 mi) border with Afghanistan in northwestern Kashmir, which is under Pakistani control.

India's border with Bangladesh runs 4,096.70 km (2,546 mi). There are 92 enclaves of Bangladesh on Indian soil and 106 enclaves of India are on Bangladeshi soil. The Teen Bigha Corridor is a strip of land formerly belonging to India on the West Bengal–Bangladesh border which has been leased indefinitely to Bangladesh so that it can access its Dehgram–Angalpota enclaves.

The Line of Actual Control (LAC) is the effective border between India and the People's Republic of China. It traverses 4,057 km along the Indian states of Jammu and Kashmir, Uttarakhand, Himachal Pradesh, Sikkim and Arunachal Pradesh. Both nations lay claim to the Aksai Chin region of northeastern Kashmir, which fell into Chinese control during the Sino-Indian War of 1962. The border with Burma (Myanmar) extends up to 1,643 km (1,021 mi) along the southern borders of India's northeastern states. Located amidst the Himalayan range, India's border with Bhutan runs 699 km (434 mi). The border with Nepal runs 1,751 km (1,088 mi) along the foothills of the Himalayas in northern India. The Siliguri Corridor, narrowed sharply by the borders of Bhutan, Nepal and Bangladesh, connects peninsular India with the northeastern states.

Physiographic Regions

India is divided into seven physiographic regions. They are;

1. The northern mountains including the Himalayas, which includes the Kuen Lun and the Karakoram ranges and the northeast mountain ranges.
2. Indo-Gangetic plains
3. Thar Desert
4. Central Highlands and Deccan Plateau
5. East Coast
6. West Coast
7. Bordering seas and islands.

Mountains

A great arc of mountains, consisting of the Himalayas, Hindu Kush, and Patkai ranges define the northern Indian subcontinent. These were formed by the ongoing tectonic collision of the Indian Plate with the Eurasian Plate that started around 50 million years ago. The mountains in these ranges include some of the world's tallest mountains which act as a natural barrier to cold polar winds. They also facilitate the monsoon winds which in turn influence the climate in India. Rivers originating in these mountains, flow through the fertile Indo–Gangetic plains. These mountains are recognised by biogeographers as the boundary between two of the Earth's great ecozones: the temperate Palearctic that covers most of Eurasia and the tropical and subtropical Indomalaya ecozone which includes the Indian subcontinent, Southeast Asia and Indonesia.

India has eight major mountain ranges having peaks of over 1,000 m (3,281 ft):

The Himalayan range is considered as the world's highest mountain range, with its tallest peak Mt. Everest on the Nepal–China border. They form India's northeastern border, separating it from northeastern Asia. They are one of the world's youngest mountain ranges and extend almost uninterrupted for 2,500 km (1,553 mi), covering an area of 500,000 km (193,051 sq mi). The Himalayas extend from Jammu and Kashmir in the west to Arunachal Pradesh in the east. These states along with Himachal Pradesh, Uttarakhand, and Sikkim lie mostly in the Himalayan region. Numerous Himalayan peaks rise over 7,000 m (22,966 ft) and the snow line ranges between 6,000 m (19,685 ft) in Sikkim to around 3,000 m (9,843 ft) in Kashmir. Kanchenjunga—on the Sikkim–Nepal border—is the highest point in the area administered by India. Most peaks in the Himalayas remain snowbound throughout the year. The Himalayas act as a barrier to the frigid katabatic winds flowing down from Central Asia. Thus, North India is kept warm or only mildly cooled during winter; in summer, the same phenomenon makes India relatively hot.

The Karakoram is situated in the disputed state of Jammu and Kashmir. It has more than sixty peaks above 7,000 m (22,966 ft), including K2, the second highest peak in the world 8,611 m (28,251 ft). K2 is just 237 m (778 ft) smaller than the 8,848 m (29,029 ft) Mount Everest. The range is about 500 km (311 mi) in length and the most heavily glaciated part of the world outside of the polar regions. The Siachen Glacier at 70 km (43 mi) and the Biafo Glacier at 63 km (39 mi) rank as the world's second and third-longest glaciers outside the polar regions. Just to the west of the northwest end of the Karakoram, lies the Hindu Raj range, beyond which is the Hindu Kush range. The southern boundary of the Karakoram is formed by the Gilgit, Indus and Shyok rivers, which separate the range from the northwestern end of the Himalayas.

The Patkai, or Purvanchal, are situated near India's eastern border with Myanmar. They were created by the same tectonic processes which led to the formation of the Himalayas. The physical features of the Patkai mountains are conical peaks, steep slopes and deep valleys. The Patkai ranges are not as rugged or tall as the Himalayas. There are three hill ranges that come under the Patkai: the Patkai–Bum, the Garo–Khasi–Jaintia and the Lushai hills. The Garo–Khasi range lies in Meghalaya. Mawsynram, a village near Cherrapunji lying on the windward side of these hills, has the distinction of being the wettest place in the world, receiving the highest annual rainfall.

The Vindhya range runs across most of central India, extending 1,050 km (652 mi). The average elevation of these hills is 3,000 m (9,843 ft). They are believed to have been formed by the wastes created by the weathering of the ancient Aravali mountains. Geographically, it separates northern India from southern India. The western end of the range lies in eastern Gujarat, near its border with Madhya Pradesh, and runs east and north, almost meeting the Ganges at Mirzapur.

The Satpura Range begins in eastern Gujarat near the Arabian Sea coast and runs east across Maharashtra, Madhya Pradesh and Chhattisgarh. It extends 900 km (559 mi) with many peaks rising above 1,000 m (3,281 ft). It is triangular in shape, with its apex at Ratnapuri and the two sides being parallel to the Tapti and Narmada rivers. It runs parallel to the Vindhya Range, which lies to the north, and these two east-west ranges divide the Indo–Gangetic plain from the Deccan Plateau located north of River Narmada. The Aravali Range is the oldest mountain range in India, running across Rajasthan from northeast to southwest direction, extending approximately 800 km (497 mi). The northern end of the range continues as isolated hills and rocky ridges into Haryana, ending near Delhi. The highest peak in this range is Guru Shikhar at Mount Abu, rising to 1,722 m (5,650 ft), lying near the border with Gujarat. The Aravali Range is the eroded stub of an ancient fold mountain system. The range rose in a Precambrian event called the Aravali–Delhi orogen. The range joins two of the ancient segments that make up the Indian craton, the Marwar segment to the northwest of the range, and the Bundelkhand segment to the southeast.

The Western Ghats or Sahyadri mountains run along the western edge of India's Deccan Plateau and separate it from a narrow coastal plain along the Arabian Sea. The range runs approximately 1,600 km (994 mi) from south of the Tapti River near the Gujarat–Maharashtra border and across Maharashtra, Goa, Karnataka, Kerala and Tamil Nadu to the southern tip of the Deccan peninsula. The average elevation is around 1,000 m (3,281 ft). Anai Mudi in the Anaimalai Hills 2,695 m (8,842 ft) in Kerala is the highest peak in the Western Ghats.

The Eastern Ghats are a discontinuous range of mountains, which have been eroded and vivisected by the four major rivers of southern India, the Godavari, Mahanadi, Krishna, and Kaveri. These mountains extend from West Bengal to Orissa, Andhra Pradesh and Tamil Nadu, along the coast and parallel to the Bay of Bengal. Though not as tall as the Western Ghats, some of its peaks are over 1,000 m (3,281 ft) in height. The Nilgiri hills in Tamil Nadu lies at the junction of the Eastern and Western Ghats.

Indo-Gangetic Plain

The Indo-Gangetic plains, also known as the *Great Plains* are large floodplains of the Indus and the Ganga-Brahmaputra river systems. They run parallel to the Himalaya mountains, from Jammu and Kashmir in the west to Assam in the east and draining most of northern and eastern India. The plains encompass an area of 700,000 square kilometers (270,272 sq mi). The major rivers in this region are the Ganges and the Indus along with their tributaries–Beas, Yamuna, Gomti, Ravi, Chambal, Sutlej and Chenab.

The great plains are sometimes classified into four divisions:

- The Bhabar belt — is adjacent to the foothills of the Himalayas and consists of boulders and pebbles which have been carried down by the river streams. As the porosity of this belt is very high, the streams flow underground. The bhabar is generally narrow with its width varying between 7 to 15 km.
- The Terai belt — lies next to the Bhabar region and is composed of newer alluvium. The underground streams reappear in this region. The region is excessively moist and thickly forested. It also receives heavy rainfall throughout the year and is populated with a variety of wildlife.
- The Bangar belt — consists of older alluvium and forms the alluvial terrace of the flood plains. In the Gangetic plains, it has a low upland covered by laterite deposits.
- The Khadar belt — lies in lowland areas after the Bangar belt. It is made up of fresh newer alluvium which is deposited by the rivers flowing down the plain.

The Indo-Gangetic belt is the world's most extensive expanse of uninterrupted alluvium formed by the deposition of silt by the numerous rivers. The plains are flat making it conducive for irrigation through canals. The area is also rich in ground water sources.

The plains are one of the world's most intensely farmed areas. The main crops grown are rice and wheat, which are grown in rotation.

Other important crops grown in the region include maize, sugarcane and cotton. The Indo-Gangetic plains rank among the world's most densely populated areas.

Thar Desert

The Thar Desert (also known as the *Great Indian Desert*) is the world's seventh largest desert; it forms a significant portion of western India and covers an area of 238,700 km^2 (92,200 mile^2). The desert continues into Pakistan as the Cholistan Desert. Most of the Thar Desert is situated in Rajasthan, covering 61% of its geographic area.

About 10 percent of this ecoregion comprises sand dunes, and the remaining 90 percent consist of craggy rock forms, compacted salt-lake bottoms, and interdunal and fixed dune areas. Annual temperatures can range from 0°C in the winter to over 50°C during the summer. Most of the rainfall received in this region is associated with the short July-September southwest monsoon that brings around 100–500 mm of precipitation. Water is scarce and occurs at great depths, ranging from 30 to 120 m below the ground level. Rainfall is precarious and erratic, ranging from below 120 mm (4.72 inches) in the extreme west to 375 mm (14.75 inches) eastward. The soils of the arid region are generally sandy to sandy-loam in texture. The consistency and depth vary as per the topographical features. The low-lying loams are heavier and may have a hard pan of clay, calcium carbonate or gypsum.

Highlands

The Central Highlands comprise of three main plateaus — the Malwa Plateau in the west, the Deccan Plateau in the south (covering most of the Indian peninsula) and the Chota Nagpur Plateau in the east.

The Malwa Plateau is spread across Rajasthan, Madhya Pradesh and Gujarat. The average elevation of the Malwa plateau is 500 metres, and the landscape generally slopes towards the north. Most of the region is drained by the Chambal River and its tributaries; the western part is drained by the upper reaches of the Mahi River. The Deccan Plateau is a large triangular plateau, bounded by the Vindhyas to the north and flanked by the Eastern and Western Ghats. The Deccan covers a total area of 1.9 million km^2 (735,000 mile^2). It is mostly flat, with elevations ranging from 300 to 600 m (1,000 to 2,000 ft). The average elevation of the plateau is 2,000 feet (600 m) above sea level. The surface slopes from 3,000 feet (900 m) in the west to 1,500 feet (450 m) in the east. It slopes gently from west to east and gives rise to several peninsular rivers such as the Godavari, the Krishna, the Kaveri and the Narmada, which drain into the Bay of Bengal. This region is mostly semi-arid as it lies on the leeward side of both Ghats. Much of the Deccan is covered by thorn scrub forest scattered with small regions of deciduous broadleaf forest. Climate in the Deccan ranges from hot summers to mild winters.

The Chota Nagpur Plateau is situated in eastern India, covering much of Jharkhand and adjacent parts of Orissa, Bihar and Chhattisgarh. Its total area is approximately 65,000 km^2 (25,000 $mile^2$) and is made up of three smaller plateaus — the Ranchi, Hazaribagh, and Kodarma plateaus. The Ranchi plateau is the largest, with an average elevation of 700 m (2,300 ft). Much of the plateau is forested, covered by the Chota Nagpur dry deciduous forests. Vast reserves of metal ores and coal have been found in the Chota Nagpur plateau. The Kathiawar peninsula in western Gujarat is bounded by the Gulf of Kutch and the Gulf of Khambat. The natural vegetation in most of the peninsula is xeric scrub, part of the Northwestern thorn scrub forests ecoregion.

In western India, the Kutch region in Gujarat and Koyna in Maharashtra are classified as a Zone IV region (high risk) for earthquakes. The Kutch city of Bhuj was the epicentre of the 2001 Gujarat earthquake, which claimed the lives of more than 20,000 people and injured 166,836 while destroying or damaging near a million homes. The 1993 Latur earthquake in Maharashtra killed 7,928 people and injured 30,000. Other areas have a moderate to low risk of an earthquake occurring.

Coasts

The Eastern Coastal Plain is a wide stretch of land lying between the Eastern Ghats and the Bay of Bengal. It stretches from Tamil Nadu in the south to West Bengal in the north. The Mahanadi, Godavari, Kaveri and Krishna rivers drain these plains and their deltas occupy most of the area. The temperature in the coastal regions exceeds 30 °C (86 °F) coupled with high levels of humidity. The region receives both the northeast and southwest monsoon rains. The southwest monsoon splits into two branches, the Bay of Bengal branch and the Arabian Sea branch. The Bay of Bengal branch moves northwards crossing northeast India in early June. The Arabian Sea branch moves northwards and discharges much of its rain on the windward side of Western Ghats. Annual rainfall in this region averages between 1,000 mm (40 in) and 3,000 mm (120 in). The width of the plains varies between 100 and 130 km (62 to 80 miles). The plains are divided into six regions — the Mahanadi delta, the southern Andhra Pradesh plain, the Krishna-Godavari deltas, the Kanyakumari coast, the Coromandel

Coast and sandy coastal. The Western Coastal Plain is a narrow strip of land sandwiched between the Western Ghats and the Arabian Sea, ranging from 50 to 100 km (30 to 60 miles) in width. It extends from Gujarat in the north and extends through Maharashtra, Goa, Karnataka and Kerala. Numerous rivers and backwaters inundate the region. Originating in the Western Ghats, the rivers are fast-flowing and mostly perennial, leading to the formation of estuaries. Major rivers flowing into the sea are the Tapi, Narmada, Mandovi and Zuari.

The coast is divided into 3 parts namely, Konkan, which is situated in Maharashtra, Goa and northern parts of Karnataka; the Kanara in Karnataka and the Malabar Coast in Kerala. Vegetation is mostly deciduous, but the Malabar Coast moist forests constitute a unique ecoregion.

Islands

The Lakshadweep and the Andaman and Nicobar Islands are India's two major island formations which are classified as union territories. The Lakshadweep Islands lie 200 to 300 km (124 to 186 miles) off the coast of Kerala in the Arabian Sea with an area of 32 km^2 (11 sq mi). They consist of 12 atolls, 3 reefs and 5 submerged banks, with a total of about 36 islands and islets. The Andaman and Nicobar Islands are located between 6° and 14° North latitude and 92° and 94° East longitude. They consist of 572 isles, lying in the Bay of Bengal near the Myanmar coast.

It is located 1255 km (780 miles) from Kolkata (Calcutta) and 193 km (120 miles) from Cape Negrais in Myanmar. The territory consists of two island groups, the Andaman Islands and the Nicobar Islands. The Andaman Islands consists of 204 small islands with a total length of 352 km (220 miles). India's only active volcano, Barren Island is situated here, having last erupted in May 2005. The Narcondum is a dormant volcano and there is a mud volcano at Baratang. Indira Point, India's southernmost land point is situated in the Nicobar islands, and lies just 189 km (117 miles) from the Indonesian island of Sumatra to the southeast. The highest point is Mount Thullier at 642 m (2,140 ft).

Significant islands just off the Indian coast include Diu, a former Portuguese enclave; Majuli, Asia's largest freshwater island; Elephanta in the Bombay Harbour; and Sriharikota barrier island in Andhra Pradesh. Salsette Island is India's most populous island on which the city of Mumbai (Bombay) is located. Forty-two islands in the Gulf of Kutch constitute the Marine National Park.

Water Bodies

India has around 14,500 km of inland navigable waterways. There are twelve rivers which are classified as major rivers, with the total catchment area exceeding 2,528,000 km^2 (976,000 sq mi). All major rivers of India originate from one of the three main watersheds:

1. The Himalaya and the Karakoram ranges
2. Vindhya and Satpura range in central India
3. Sahyadri or Western Ghats in western India.

The Himalayan river networks are snow-fed and have a perennial supply throughout the year. The other two river systems are dependent on the monsoons and shrink into rivulets during the dry season. The Himalayan rivers that flow westward into Pakistan are the Indus, Beas, Chenab, Ravi, Sutlej, and Jhelum.

The Ganga-Brahmaputra-Meghana system has the largest catchment area of 1,100,000 km^2 (420,000 sq mi). The Ganga originates from the Gangotri Glacier in Uttarakhand. It flows southeast, draining into the Bay of Bengal. The Yamuna and Gomti rivers also arise in the western Himalayas and join the Ganga in the plains. The Brahmaputra, another tributary of the Ganga, originates in Tibet and enters India through the far-eastern state of Arunachal Pradesh. It proceeds westwards, joining the Ganges in Bangladesh.

The Chambal, another tributary of the Ganga originates from the Vindhya-Satpura watershed. The river flows eastward. Westward-flowing rivers from this watershed are the Narmada and Tapti, which drain into the Arabian Sea in Gujarat. The river network that flows from east to west constitutes 10% of the total outflow.

The Western Ghats are the source of all Deccan rivers, which include the Mahanadi River through the Mahanadi River Delta, Godavari River, Krishna River and Kaveri River, all draining into the Bay of Bengal. These rivers constitute 20% of India's total outflow. The heavy southwest monsoon rains cause the Brahmaputra and other rivers to distend their banks, often flooding surrounding areas. Though they provide rice paddy farmers with a largely dependable source of natural irrigation and fertilisation, such floods have killed thousands of people and tend to cause displacements of people in such areas.

Major gulfs include the Gulf of Cambay, Gulf of Kutch and the Gulf of Mannar. Straits include the Palk Strait, which separates India from Sri Lanka and the Ten Degree Channel, which separates the Andamans from the Nicobar Islands and the Eight Degree Channel, which separates the Laccadive and Amindivi Islands from the Minicoy

Island towards the south. Important capes include the Kanyakumari, the southern tip of mainland India; Indira Point, the southernmost location of India; Rama's Bridge and Point Calimere. While, Arabian Sea lies on the western side of India, Bay of Bengal and Indian Ocean lie towards the eastern and southern side respectively. Smaller seas include the Laccadive Sea and the Andaman Sea. There are four coral reefs in India, located in the Andaman and Nicobar Islands, Gulf of Mannar, Lakshadweep and Gulf of Kutch. Important lakes include Chilka Lake, the country's largest saltwater lake in Orissa; Kolleru Lake in Andhra Pradesh; Loktak Lake in Manipur, Dal Lake in Kashmir, Sambhar Lake in Rajasthan and the Sasthamkotta Lake in Kerala.

Wetlands

India's wetland ecosystem is widely distributed from the cold and arid located in the Ladakh region of Jammu and Kashmir, and those with the wet and humid climate of peninsular India. Most of the wetlands are directly or indirectly linked to river networks. The Indian government has identified a total of 71 wetlands for conservation and are part of sanctuaries and national parks. Mangrove forests are present all along the Indian coastline in sheltered estuaries, creeks, backwaters, salt marshes and mudflats. The mangrove area covers a total of 4,461 km^2 (1,722 sq mi), which comprises 7% of the world's total mangrove cover. Prominent mangrove covers are located in the Andaman and Nicobar Islands, the Sundarbans delta, the Gulf of Kutch and the deltas of the Mahanadi, Godavari and Krishna rivers. Parts of Maharashtra, Karnataka and Kerala also have large mangrove covers.

The Sundarbans delta is home to the largest mangrove forest in the world. It lies at the mouth of the Ganges and spreads across areas of Bangladesh and West Bengal. The Sundarbans is a UNESCO World Heritage Site, but is identified separately as the Sundarbans (Bangladesh) and the Sundarbans National Park (India). The Sundarbans are intersected by a complex network of tidal waterways, mudflats and small islands of salt-tolerant mangrove forests. The area is known for its diverse fauna, being home to a large variety of species of birds, spotted deer, crocodiles and snakes. Its most famous inhabitant is the Bengal Tiger. It is estimated that there are now 400 Bengal tigers and about 30,000 spotted deer in the area.

The Rann of Kutch is a marshy region located in northwestern Gujarat and the bordering Sindh province of Pakistan. It occupies a total area of 27 900 km^2 (10,800 mile^2). The region was originally a part of the Arabian Sea. Geologic forces such as earthquakes resulted in the damming up of the region, turning it into a large saltwater lagoon.

This area gradually filled with silt thus turning it into a seasonal salt marsh. During the monsoons, the area turns into a shallow marsh, often flooding to knee-depth. After the monsoons, the region turns dry and becomes parched.

Climate

Based on the Koppen system, India hosts six major climatic subtypes, ranging from arid desert in the west, alpine tundra and glaciers in the north, and humid tropical regions supporting rainforests in the southwest and the island territories. Many regions have starkly different microclimates. The nation has four seasons: winter (January–February), summer (March–May), a monsoon (rainy) season (June–September) and a post-monsoon period (October–December).

The Himalayas act as a barrier to the frigid katabatic winds flowing down from Central Asia. Thus, North India is kept warm or only mildly cooled during winter; in summer, the same phenomenon makes India relatively hot. Although the Tropic of Cancer—the boundary between the tropics and subtropics—passes through the middle of India, the whole country is considered to be tropical.

Summer lasts between March and June in most parts of India. Temperatures exceed 40 °C (104 °F) during the day. The coastal regions exceed 30 °C (86 °F) coupled with high levels of humidity. In the Thar desert area temperatures can exceed 45 °C (113 °F). The rain-bearing monsoon clouds are attracted to the low-pressure system created by the Thar Desert. The southwest monsoon splits into two arms, the Bay of Bengal arm and the Arabian Sea arm. The Bay of Bengal arm moves northwards crossing northeast India in early June. The Arabian Sea arm moves northwards and deposits much of its rain on the windward side of Western Ghats. Winters in peninsula India see mild to warm days and cool nights. Further north the temperature is cooler. Temperatures in some parts of the Indian plains sometimes fall below freezing. Most of northern India is plagued by fog during this season. The highest temperature recorded in India was 50.6 °C (123.1 °F) in Alwar in 1955. The lowest was “45 °C (“49.0 °F) in Kashmir.

Geology

India's geological features are classified based on their era of formation. The Precambrian formations of Cudappah and Vindhyan systems are spread out over the eastern and southern states. A small part of this period is spread over western and central India. The Paleozoic formations from the Cambrian, Ordovician, Silurian and Devonian system are found in the Western Himalaya region in Kashmir and Himachal Pradesh. The Mesozoic Deccan Traps formation is seen over most of the northern Deccan; they are believed to be the result of sub-aerial volcanic activity. The Trap soil is black in colour and conducive to agriculture. The Carboniferous system, Permian System and Triassic systems are seen in the western Himalayas. The Jurassic system is seen in the western Himalayas and Rajasthan.

Tertiary imprints are seen in parts of Manipur, Nagaland, Arunachal Pradesh and along the Himalayan belt. The Cretaceous system is seen in central India in the Vindhyas and part of the Indo-Gangetic plains. The Gondowana system is seen in the Narmada River area in the Vindhyas and Satpuras. The Eocene system is seen in the western Himalayas and Assam. Oligocene formations are seen in Kutch and Assam. The Pleistocene system is found over central India. The Andaman and Nicobar Island are thought to have been formed in this era by volcanoes. The Himalayas were formed by the convergence and deformation of the Indo-Australian and Eurasian Plates. Their continued convergence raises the height of the Himalayas by 1 cm each year.

Soils in India can be classified into 8 categories: alluvial, black, red, laterite, forest, arid & desert, saline & alkaline and peaty & organic soils. Alluvial soil constitute the largest soil group in India, constituting 80% of the total land surface. It is derived from the deposition of silt carried by rivers and are found in the Great Northern plains from Punjab to the Assam valley. Alluvial soil are generally fertile but they lack nitrogen and tend to bc phosphoric.

Black soil are well developed in the Deccan lava region of Maharashtra, Gujarat, and Madhya Pradesh. These contain high percentage of clay and are moisture retentive. Red soil are found in Tamil Nadu, Karnataka plateau, Andhra plateau, Chota Nagpur plateau and the Aravallis. These are deficient in nitrogen, phosphorus and humus. Laterite soils are formed in tropical regions with heavy rainfall. Heavy rainfall results in leaching out all soluble material of top layer of soil. These are generally found in Western ghats, Eastern ghats and hilly areas of northeastern states that receive heavy rainfall. Forest soils occur on the slopes of mountains and hills in Himalayas, Western Ghats and Eastern Ghats. These generally consist of large amounts of dead leaves and other organic matter called humus.

Natural Resources

Indian coal production is the 3rd highest in the world according to the 2008 Indian Ministry of Mines estimates. Shown above is a coal mine in Jharkhand.

India's total renewable water resources are estimated at 1,907.8 km3/year. Its annual supply of usable and replenshable groundwater amounts to 350 billion cubic metres. Only 35% of groundwater resources are being utilised. About 44 million tonnes of cargo is moved annually through the country's major rivers and waterways. Groundwater supplies 40% of water in India's irrigation canals. 56% of the land is arable and used for agriculture. Black soils are moisture-retentive and are preferred for dry farming and growing cotton, linseed, etc. Forest soils are used for tea and coffee plantations. Red soil have a wide diffusion of iron content.

Most of India's estimated 5.4 billion barrels (860,000,000 m^3) in oil reserves are located in the Mumbai High, upper Assam, Cambay, the Krishna-Godavari and Cauvery basins. India possesses about seventeen trillion cubic feet of natural gas in Andhra Pradesh, Gujarat and Orissa. Uranium is mined in Andhra Pradesh. India has 400 medium-to-high enthalpy thermal springs for producing geothermal energy in seven "provinces" — the Himalayas, Sohana, Cambay, the Narmada-Tapti delta, the Godavari delta and the Andaman and Nicobar Islands (specifically the volcanic Barren Island.)

India is the world's biggest producer of mica blocks and mica splittings. India ranks second amongst the world's largest producers of barites and chromites. The Pleistocene system is rich in minerals. India is the third-largest coal producer in the world and ranks fourth in the production of iron ore. It is the fifth-largest producer of bauxite and crude steel, the seventh-largest of manganese ore and the eighth-largest of aluminium. India has significant sources of titanium ore, diamonds and limestone. India possesses 24% of the world's known and economically-viable thorium, which is mined along shores of Kerala. Gold had been mined in the now-defunct Kolar Gold Fields in Karnataka.

Flora and Fauna

The Bengal tiger is the national animal of India. India is home to about half of the world's tiger population but the future of the species is threatened by habitat degradation and poaching.

India, which lies within the Indomalaya ecozone, displays significant biodiversity. One of eighteen megadiverse countries, it is home to 7.6% of all mammalian, 12.6% of all avian, 6.2% of all reptilian, 4.4% of all amphibian, 11.7% of all fish, and 6.0% of all flowering plant species. Many ecoregions, such as the *shola* forests, exhibit extremely high rates of endemism; overall, 33% of Indian plant species are endemic.

India's forest cover ranges from the tropical rainforest of the Andaman Islands, Western Ghats, and North-East India to the coniferous forest of the Himalaya. Between these extremes lie the sal-dominated moist deciduous forest of eastern India; the teak-dominated dry deciduous forest of central and southern India; and the babul-dominated thorn forest of the central Deccan and western Gangetic plain. Important Indian trees include the medicinal neem, widely used in rural Indian herbal remedies. The pipal fig tree, shown on the seals of Mohenjo-daro, shaded Gautama Buddha as he sought enlightenment. According to latest report, less than 12% of India's landmass is covered by dense forests.

Many Indian species are descendants of taxa originating in Gondwana, from which the Indian plate separated. Peninsular India's subsequent movement towards, and collision with, the Laurasian landmass set off a mass exchange of species. However, volcanism and climatic changes 20 million years ago caused the extinction of many endemic Indian forms. Soon thereafter, mammals entered India from Asia through two zoogeographical passes on either side of the emerging Himalaya. Consequently, among Indian species, only 12.6% of mammals and 4.5% of birds are endemic, contrasting with 45.8% of reptiles and 55.8% of amphibians. Notable endemics are the Nilgiri leaf monkey and the brown and carmine Beddome's toad of the Western Ghats. India contains 172, or 2.9%, of IUCN-designated threatened species. These include the Asiatic Lion, the Bengal Tiger, and the Indian white-rumped vulture, which suffered a near-extinction from ingesting the carrion of diclofenac-treated cattle.

In recent decades, human encroachment has posed a threat to India's wildlife; in response, the system of national parks and protected areas, first established in 1935, was substantially expanded. In 1972, India enacted the Wildlife Protection Act and Project Tiger to safeguard crucial habitat; in addition, the Forest Conservation Act was enacted in 1980. Along with more than five hundred wildlife sanctuaries, India hosts thirteen biosphere reserves, four of which are part of the World Network of Biosphere Reserves; twenty-five wetlands are registered under the Ramsar Convention.

Economy

The economy of India is the eleventh largest economy in the world by nominal GDP (just behind Canada, Spain and Brazil) and the fourth largest by purchasing power parity (PPP). Following strong economic reforms from the socialist inspired economy of a post-independence Indian nation, the country began to develop a fast-paced economic growth, as markets opened for international competition and investment, in the 1990s. In the 21st century, India is an emerging economic power with vast human and natural resources, and a huge knowledge base. Economists predict that by 2020, India will be among the leading economies of the world.

India was under social democratic-based policies from 1947 to 1991. The economy was characterised by extensive regulation, protectionism, public ownership, pervasive corruption and slow growth. Since 1991, continuing economic liberalisation has moved the economy towards a market-based system. A revival of economic reforms and better economic policy in 2000s accelerated India's economic growth rate. In recent years, Indian cities have continued to liberalize business regulations. By 2008, India had established itself as the world's second-fastest growing major economy. However, the year 2009 saw a significant slowdown in India's GDP growth rate to 6.8% as well as the return of a large projected fiscal deficit of 6.8% of GDP which would be among the highest in the world.

India's large service industry accounts for 62.5% of the country's Gross Domestic Product (GDP) while the industrial and agricultural sector contribute 20% and 17.5% respectively. Agriculture is the predominant occupation in India, accounting for about 52% of employment. The service sector makes up a further 34%, and industrial sector around 14%. The labour force totals half a billion workers. Major agricultural products include rice, wheat, oilseed, cotton, jute, tea, sugarcane, potatoes, cattle, water buffalo, sheep, goats, poultry and fish. Major industries include telecommunications, textiles, chemicals, food processing, steel, transportation equipment, cement, mining, petroleum, machinery, information technology enabled services and software.

India's per capita income (nominal) is $1,030, ranked 139th in the world, while its per capita (PPP) of US$2,940 is ranked 128th. Previously a closed economy, India's trade has grown fast. India currently accounts for 1.5% of World trade as of 2007 according to the WTO. According to the World Trade Statistics of the WTO in 2006, India's total merchandise trade (counting exports and imports) was valued at $294 billion in 2006 and India's services trade inclusive of export and import was $143 billion. Thus, India's global economic engagement in 2006 covering both merchandise and services trade was of the order of $437 billion, up by a record 72% from a level of $253 billion in 2004. India's trade has reached a still relatively moderate share 24% of GDP in 2006, up from 6% in 1985.

History

India's economic history can be broadly divided into three eras, beginning with the pre-colonial period lasting up to the 18th century. The advent of British colonisation started the colonial period in the early 19th century, which ended with independence in 1947. The third period stretches from independence in 1947 until now.

Pre-colonial

The citizens of the Indus Valley civilisation, a permanent settlement that flourished between 2800 BC and 1800 BC, practiced agriculture, domesticated animals, used uniform weights and measures, made tools and weapons, and traded with other cities. Evidence of well planned streets, a drainage system and water supply reveals their knowledge of urban planning, which included the world's first urban sanitation systems and the existence of a form of municipal government.

The 1872 census revealed that 99.3% of the population of the region constituting present-day India resided in villages, whose economies were largely isolated and self-sustaining, with agriculture the predominant occupation. This satisfied the food requirements of the village and provided raw materials for hand-based industries, such as textiles, food processing and crafts. Although many kingdoms and rulers issued coins, barter was prevalent. Villages paid a portion of their agricultural produce as revenue to the rulers, while its craftsmen received a part of the crops at harvest time for their services.

Religion, especially Hinduism, and the caste and the joint family systems, played an influential role in shaping economic activities. The caste system functioned much like medieval European guilds, ensuring the division of labour, providing for the training of apprentices and, in some cases, allowing manufacturers to achieve narrow specialization. For instance, in certain regions, producing each variety of cloth was the specialty of a particular sub-caste.

Textiles such as muslin, Calicos, shawls, and agricultural products such as pepper, cinnamon, opium and indigo were exported to Europe, the Middle East and South East Asia in return for gold and silver.

Assessment of India's pre-colonial economy is mostly qualitative, owing to the lack of quantitative information. One estimate puts the revenue of Akbar's Mughal Empire in 1600 at £17.5 million, in contrast with the total revenue of Great Britain in 1800, which totalled £16 million. India, by the time of the arrival of the British, was a largely traditional agrarian economy with a dominant subsistence sector dependent on primitive technology. It existed alongside a competitively developed network of commerce, manufacturing and credit. After the decline of the Mughals, western, central and parts of south and north India were integrated and administered by the Maratha Empire. The Maratha Empire's budget in 1740s, at its peak, was Rs. 100 million. After the loss at Panipat, the Maratha Empire disintegrated into confederate states of Gwalior, Baroda, Indore, Jhansi, Nagpur, Pune and Kolhapur. Gwalior state had a budget of Rs. 30M. However, at this time, British East India company entered the Indian political theatre. Until 1857, when India was firmly under the British crown, the country remained in a state of political instability due to internecine wars and conflicts.

Colonial

Company rule in India brought a major change in the taxation environment from revenue taxes to property taxes, resulting in mass impoverishment and destitution of majority of farmers and led to numerous famines. The economic policies of the British Raj effectively bankrupted India's large handicrafts industry and caused a massive drain of India's resources. Indian Nationalists employed the successful Swadeshi movement, as strategy to diminish British economic superiority by boycotting British products and the reviving the market for domestic-made products and production techniques. India had become a strong market for superior finished European goods. This was because of vast gains made by the Industrial revolution in Europe, the effects of which was deprived to Colonial India.

The Nationalists had hoped to revive the domestic industries that were badly effected by polices implemented by British Raj which had made them uncompetitive to British made goods.

An estimate by Cambridge University historian Angus Maddison reveals that "India's share of the world income fell from 22.6% in 1700, comparable to Europe's share of 23.3%, to a low of 3.8% in 1952". It also created an institutional environment that, on paper, guaranteed property rights among the colonizers, encouraged free trade, and created a single currency with fixed exchange rates, standardized weights and measures, capital markets. It also established a well developed system of railways and telegraphs, a civil service that aimed to be free from political interference, a common-law and an adversarial legal system. India's colonisation by the British coincided with major changes in the world economy—industrialisation, and significant growth in production and trade. However, at the end of colonial rule, India inherited an economy that was one of the poorest in the developing world, with industrial development stalled, agriculture unable to feed a rapidly growing population, India had one of the world's lowest life expectancies, and low rates for literacy. The impact of the British rule on India's economy is a controversial topic. Leaders of the Indian independence movement, and left-nationalist economic historians have blamed colonial rule for the dismal state of India's economy in its aftermath and that financial strength required for Industrial development in Europe was derived from the wealth taken from Colonies in Asia and Africa. At the same time right-wing historians have countered that India's low economic performance was due to various sectors being in a state of growth and decline due to changes brought in by colonialism and a world that was moving towards industrialization and economic integration.

Independence to 1991

Indian economic policy after independence was influenced by the colonial experience (which was seen by Indian leaders as exploitative in nature) and by those leaders' exposure to Fabian socialism. Policy tended towards protectionism, with a strong emphasis on import substitution, industrialization, state intervention in labour and financial markets, a large public sector, business regulation, and central planning. Five-Year Plans of India resembled central planning in the Soviet Union. Steel, mining, machine tools, water, telecommunications, insurance, and electrical plants, among other industries, were effectively nationalized in the mid-1950s. Elaborate licences, regulations and the accompanying red tape, commonly referred to as Licence Raj, were required to set up business in India between 1947 and 1990.

Jawaharlal Nehru, the first prime minister, along with the statistician Prasanta Chandra Mahalanobis, carried on by Indira Gandhi formulated and oversaw economic policy. They expected favourable outcomes from this strategy, because it involved both public and private sectors and was based on direct and indirect state intervention, rather than the more extreme Soviet-style central command system. The policy of concentrating simultaneously on capital-and technology-intensive heavy industry and subsidizing manual, low-skill cottage industries was criticized by economist Milton Friedman, who thought it would waste capital and labour, and retard the development of small manufacturers. The rate from 1947–80 was derisively referred to as the Hindu rate of growth, because of the unfavourable comparison with growth rates in other Asian countries, especially the "East Asian Tigers".

The Rockefeller Foundation's research in high-yielding varieties of seeds, their introduction after 1965 and the increased use of fertilizers and irrigation are known collectively as the Green Revolution in India, which provided the increase in production needed to make India self-sufficient in food grains, thus improving agriculture in India. Famine in India, once accepted as inevitable, has not returned since independence.

Since 1991

In the late 80s, the government led by Rajiv Gandhi eased restrictions on capacity expansion for incumbents, removed price controls and reduced corporate taxes. While this increased the rate of growth, it also led to high fiscal deficits and a worsening current account. The collapse of the Soviet Union, which was India's major trading partner, and the first Gulf War, which caused a spike in oil prices, caused a major balance-of-payments crisis for India, which found itself facing the prospect of defaulting on its loans. India asked for a $1.8 billion bailout loan from IMF, which in return demanded reforms.

In response, Prime Minister Narasimha Rao along with his finance minister and current Prime Minister of India Dr. Manmohan Singh initiated the economic liberalisation of 1991. The reforms did away with the Licence Raj (investment, industrial and import licensing) and ended many public monopolies, allowing automatic approval of foreign direct investment in many sectors.

Since then, the overall direction of liberalisation has remained the same, irrespective of the ruling party, although no party has tried to take on powerful lobbies such as the trade unions and farmers, or contentious issues such as reforming labour laws and reducing agricultural subsidies. Since 1990 India has emerged as one of the fastest-growing economies in the developing world; during this period, the economy has grown constantly, but with a few major setbacks. This has been accompanied by increases in life expectancy, literacy rates and food security.

While the credit rating of India was hit by its nuclear tests in 1998, it has been raised to investment level in 2007 by S&P and Moody's. In 2003, Goldman Sachs predicted that India's GDP in current prices will overtake France and Italy by 2020, Germany, UK and Russia by 2025 and Japan by 2035. By 2035, it was projected to be the third largest economy of the world, behind US and China. India is often seen by most economists as a rising economic superpower and is believed to play a major role in the global economy in the 21st century. In 2009 India purchased 200 Tons of Gold for $6.7 billion from IMF as a total role reversal from 1991.

Industry and Services

Industry accounts for 54.6% of the GDP and employ 17% of the total workforce. However, about one-third of the industrial labour force is engaged in simple household manufacturing only. In absolute terms, India is 16th in the world in terms of nominal factory output..

Economic reforms brought foreign competition, led to privatisation of certain public sector industries, opened up sectors hitherto reserved for the public sector and led to an expansion in the production of fast-moving consumer goods. Post-liberalisation, the Indian private sector, which was usually run by oligopolies of old family firms and required political connections to prosper was faced with foreign competition, including the threat of cheaper Chinese imports. It has since handled the change by squeezing costs, revamping management, focusing on designing new products and relying on low labour costs and technology.

Textile manufacturing is the second largest source for employment after agriculture and accounts for 26% of manufacturing output. Ludhiana produces 90% of woolens in India and is also Known as the Manchester of India. Tirupur has gained universal recognition as the leading source of hosiery, knitted garments, casual wear and sportswear. Dharavi slum in Mumbai has gained fame for leather products. Tata Motors' Nano attempts to be the world's cheapest car.

India is fifteenth in services output. It provides employment to 23% of work force, and it is growing fast, growth rate 7.5% in 1991–2000 up from 4.5% in 1951–80. It has the largest share in the GDP, accounting for 55% in 2007 up from 15% in 1950. Business services (information technology, information technology enabled services, business process outsourcing) are among the fastest growing sectors contributing to one third of the total output of services in 2000. The growth in the IT sector is attributed to increased specialization, and an availability of a large pool of low cost, but highly skilled, educated and fluent English-speaking workers, on the supply side, matched on the demand side by an increased demand from foreign consumers interested in India's service exports, or those looking to outsource their operations. The share of India's IT industry to the country's GDP increased from 4.8 % in 2005-06 to 7% in 2008. In 2009, seven Indian firms were listed among the top 15 technology outsourcing companies in the world. In March 2009, annual revenues from outsourcing operations in India amounted to US$60 billion and this is expected to increase to US$225 billion by 2020.

Organized retail such supermarkets accounts for 24% of the market as of 2008. Regulations prevent most foreign investment in retailing.

Moreover, over thirty regulations such as "signboard licences" and "anti-hoarding measures" may have to be complied before a store can open doors. There are taxes for moving goods to states, from states, and even within states.

Tourism in India is relatively undeveloped, but growing at double digits. Some hospitals woo medical tourism.

Agriculture

India ranks second worldwide in farm output. Agriculture and allied sectors like forestry, logging and fishing accounted for 16.6% of the GDP in 2007, employed 60% of the total workforce and despite a steady decline of its share in the GDP, is still the largest economic sector and plays a significant role in the overall socio-economic development of India. Yields per unit area of all crops have grown since 1950, due to the special emphasis placed on agriculture in the five-year plans and steady improvements in irrigation, technology, application of modern agricultural practices and provision of agricultural credit and subsidies since Green revolution in India. However, international comparisons reveal the average yield in India is generally 30% to 50% of the highest average yield in the world.

India is the largest producer in the world of milk, cashew nuts, coconuts, tea, ginger, turmeric and black pepper. It also has the world's largest cattle population: 193 million. It is the second largest producer of wheat, rice, sugar, cotton, silk, peanuts and inland fish. It is the third largest producer of tobacco. India is the largest fruit producer, accounting for 10% of the world fruit production. It is the leading producer of bananas, sapotas and mangoes.

India is the second largest producer and the largest consumer of silk in the world, with the majority of the 77 million kg (2005) production taking place in Karnataka State, particularly in Mysore and the North Bangalore regions of Muddenahalli, Kanivenarayanapura, and Doddaballapura, the upcoming sites of a INR 700 million "Silk City".

Banking and Finance

The Indian money market is classified into: the organised sector (comprising private, public and foreign owned commercial banks and cooperative banks, together known as *scheduled banks*); and the unorganised sector (comprising individual or family owned indigenous bankers or money lenders and non-banking financial companies (NBFCs)). The unorganised sector and microcredit are still preferred over traditional banks in rural and sub-urban areas, especially for non-productive purposes, like ceremonies and short duration loans.

Prime Minister Indira Gandhi nationalised 14 banks in 1969, followed by six others in 1980, and made it mandatory for banks to provide 40% of their net credit to priority sectors like agriculture, small-scale industry, retail trade, small businesses, etc. to ensure that the banks fulfill their social and developmental goals. Since then, the number of bank branches has increased from 10,120 in 1969 to 98,910 in 2003 and the population covered by a branch decreased from 63,800 to 15,000 during the same period. The total deposits increased 32.6 times between 1971 to 1991 compared to 7 times between 1951 to 1971. Despite an increase of rural branches, from 1,860 or 22% of the total number of branches in 1969 to 32,270 or 48%, only 32,270 out of 5 lakh (500,000) villages are covered by a scheduled bank. The public sector banks hold over 75% of total assets of the banking industry, with the private and foreign banks holding 18.2% and 6.5% respectively. Since liberalisation, the government has approved significant banking reforms. While some of these relate to nationalised banks (like encouraging mergers, reducing government interference and increasing profitability and competitiveness), other reforms have opened up the banking and insurance sectors to private and foreign players. More than half of personal savings are invested in physical assets such as land, houses, cattle, and gold.

Natural Resources

India's total cultivable area is 1,269,219 km² (56.78% of total land area), which is decreasing due to constant pressure from an ever growing population and increased urbanisation. India has a total water surface area of 314,400 km² and receives an average annual rainfall of 1,100 mm. Irrigation accounts for 92% of the water utilisation, and comprised 380 km² in 1974, and is expected to rise to 1,050 km² by 2025, with the balance accounted for by industrial and domestic consumers. India's inland water resources comprising rivers, canals, ponds and lakes and marine resources comprising the east and west coasts of the Indian ocean and other gulfs and bays provide employment to nearly 6 million people in the fisheries sector. In 2008, India had the world's third largest fishing industry.

India's major mineral resources include coal, iron, manganese, mica, bauxite, titanium, chromite, limestone and thorium. India meets most of its domestic energy demand through its 92 billion tonnes of coal reserves (about 10% of world's coal reserves).

India's huge thorium reserves — about 25% of world's reserves — is expected to fuel the country's ambitious nuclear energy program in the long-run. India's dwindling uranium reserves stagnated the growth of nuclear energy in the country for many years. However, the Indo-US nuclear deal has paved the way for India to import uranium from other countries. India is also believed to be rich in certain renewable sources of energy with significant future potential such as solar, wind and biofuels (jatropha, sugarcane).

Petroleum and Natural Gas

India's oil reserves, found in Bombay High, parts of Gujarat, Rajasthan and eastern Assam, meet 25% of the country's domestic oil demand. India's total proven oil reserves stand at 11 billion barrels, of which Bombay High is believed to hold 6.1 billion barrels and Mangala Area in Rajasthan an additional 3.6 billion barrels.

In 2009, India imported 2.56 million barrels of oil per day, making it one of largest buyers of crude oil in the world. The petroleum industry in India mostly consists of public sector companies such as Oil and Natural Gas Corporation (ONGC), Hindustan Petroleum Corporation Limited (HPCL) and Indian Oil Corporation Limited (IOCL). There are some major private Indian companies in oil sector such as Reliance Industries Limited (RIL) which operates the world's largest oil refining complex.

Pharmaceuticals

India has a self reliant Pharmaceuticals industry. The majority of its medical consumables are produced domestically. Pharmaceutical Industry in India is dotted with companies like Ranbaxy Laboratories, Dr. Reddy's Laboratories, Cipla which have created a niche for themselves at world level.

Today, India is an exporter to countries like the United States and Russia. In terms of the global market, India currently holds a modest 1-2% share, but it has been growing at approximately 10% per year. Indian Pharmaceutical Industry is often compared to Pharmaceutical Industry in the USA.

External Trade and Investment

Global Trade Relations

India's economy is mostly dependent on its large internal market with external trade accounting for just 20% of the country's GDP. In 2008, India accounted for 1.45% of global merchandise trade and 2.8% of global commercial services export. Until the liberalization of 1991, India was largely and intentionally isolated from the world markets, to protect its economy and to achieve self-reliance. Foreign trade was subject to import tariffs, export taxes and quantitative restrictions, while foreign direct investment (FDI) was restricted by upper-limit equity participation, restrictions on technology transfer, export obligations and government approvals; these approvals were needed for nearly 60% of new FDI in the industrial sector.

The restrictions ensured that FDI averaged only around US$200 million annually between 1985 and 1991; a large percentage of the capital flows consisted of foreign aid, commercial borrowing and deposits of non-resident Indians. India's exports were stagnant for the first 15 years after independence, due to the predominance of tea, jute and cotton manufactures, demand for which was generally inelastic. Imports in the same period consisted predominantly of machinery, equipment and raw materials, due to nascent industrialization.

Since liberalization, the value of India's international trade has become more broad-based and has risen to Rs. 63,080,109 crores in 2003–04 from Rs.1,250 crores in 1950–51. India's major trading partners are China, the US, the UAE, the UK, Japan and the EU. The exports during April 2007 were $12.31 billion up by 16% and import were $17.68 billion with an increase of 18.06% over the previous year. In 2006-07, major export commodities included engineering goods, petroleum products, chemicals and pharmaceuticals, gems and jewellery, textiles and garments, agricultural products, iron ore and other minerals. Major import commodities included crude oil and related products, machinery, electronic goods, gold and silver.

India is a founding-member of General Agreement on Tariffs and Trade (GATT) since 1947 and its successor, the WTO. While participating actively in its general council meetings, India has been crucial in voicing the concerns of the developing world. For instance, India has continued its opposition to the inclusion of such matters as labour and environment issues and other *non-tariff barriers* into the WTO policies.

Balance of Payments

Since independence, India's balance of payments on its current account has been negative. Since liberalisation in the 1990s (precipitated by a balance of payment crisis), India's exports have been consistently rising, covering 80.3% of its imports in 2002–03, up from 66.2% in 1990–91. India's growing oil import bill is seen as the main driver behind the large current account deficit. In 2007-08, India imported 120.1 million tonnes of crude oil, more than 3/4th of the domestic demand, at a cost of $61.72 billion.

Although India is still a net importer, since 1996–97 its overall balance of payments (i.e., including the capital account balance) has been positive, largely on account of increased foreign direct investment and deposits from non-resident Indians; until this time, the overall balance was only occasionally positive on account of external assistance and commercial borrowings. As a result, India's foreign currency reserves stood at $285 billion in 2008.

Due to the global late-2000s recession, both Indian exports and imports declined by 29.2% and 39.2% respectively in June 2009. The steep decline was because countries hit hardest by the global recession, such as United States and members of the European Union, account for more than 60% of Indian exports. However, since the decline in imports was much sharper compared to the decline in exports, India's trade deficit reduced to 252.5 billion rupee.

India's reliance on external assistance and commercial borrowings has decreased since 1991–92, and since 2002–03, it has gradually been repaying these debts. Declining interest rates and reduced borrowings decreased India's debt service ratio to 4.5% in 2007. In India, External Commercial Borrowings (ECBs) are being permitted by the Government for providing an additional source of funds to Indian corporates. The Ministry of Finance monitors and regulates these borrowings (ECBs) through ECB policy guidelines.

Foreign Direct Investment in India

As the fourth-largest economy in the world in PPP terms, India is a preferred destination for foreign direct investments (FDI); India has strengths in telecommunication, information technology and other significant areas such as auto components, chemicals, apparels, pharmaceuticals, and jewellery. Despite a surge in foreign investments, rigid FDI policies resulted in a significant hindrance. However, due to some positive economic reforms aimed at deregulating the economy and stimulating foreign investment, India has positioned itself as one of the front-runners of the rapidly growing Asia Pacific Region. India has a large pool of skilled managerial and technical expertise. The size of the middle-class population stands at 300 million and represents a growing consumer market. The inordinately high investment from Mauritius is due to routing of international funds through the country given significant capital gains tax advantages; double taxation is avoided due to a tax treaty between India and Mauritius, and Mauriitus is a capital gains tax haven, effectively creating a zero-taxation FDI channel.

India's recently liberalized FDI policy (2005) allows up to a 100% FDI stake in ventures. Industrial policy reforms have substantially reduced industrial licensing requirements, removed restrictions on expansion and facilitated easy access to foreign technology and foreign direct investment FDI. The upward moving growth curve of the real-estate sector owes some credit to a booming economy and liberalized FDI regime. In March 2005, the government amended the rules to allow 100 per cent FDI in the construction business. This automatic route has been permitted in townships, housing, built-up infrastructure and construction development projects including housing, commercial premises, hotels, resorts, hospitals, educational institutions, recreational facilities, and city-and regional-level infrastructure.

A number of changes were approved on the FDI policy to remove the caps in most sectors. Fields which require relaxation in FDI restrictions include civil aviation, construction development, industrial parks, petroleum and natural gas, commodity exchanges, credit-information services and mining. But this still leaves an unfinished agenda of permitting greater foreign investment in politically sensitive areas such as insurance and retailing. FDI inflows into India reached a record $19.5 billion in fiscal year 2006-07 (April-March), according to the government's Secretariat for Industrial Assistance. This was more than double the total of US$7.8bn in the previous fiscal year. The FDI inflow for 2007-08 has been reported as $24 billion and for 2008-09, it is expected to be above $35 billion. A critical factor in determining India's continued economic growth and realizing the potential to be an economic superpower is going to depend on how the government can create incentives for FDI flow across a large number of sectors in India.

Currency

The Indian rupee is the only legal tender accepted in India. The exchange rate as on 23 March 2010 is 45.40 INR the USD, 61.45 to a EUR, and 68.19 to a GBP. The Indian rupee is accepted as legal tender in the neighbouring Nepal and Bhutan, both of which peg their currency to that of the Indian rupee. The rupee is divided into 100 paise. The highest-denomination banknote is the 1,000 rupee note; the lowest-denomination coin in circulation is the 25 paise coin (it earlier had 1, 2, 5, 10 and 20 paise coins which have been discontinued by the Reserve Bank of India).

The Rupee hit a record low during early 2009 on account of global recession. However, due to a strong domestic market, India managed to bounce back sooner than the western countries. Since September

2009 there has been a constant appreciation in Rupee versus most Tier 1 currencies. On 11 January 2010 Rupee went as high as 45.50 to a United states dollar and on 10 January 2010 as high as Rupee 73.93 to a British Pound. A rising rupee also prompted Government of India to buy 200 tonnes of Gold from IMF.

The RBI, the country's central bank was established on 1 April 1935. It serves as the nation's monetary authority, regulator and supervisor of the financial system, manager of exchange control and as an issuer of currency. The RBI is governed by a central board, headed by a governor who is appointed by the Central government of India.

Income and Consumption

As of 2005:

- 85.7% of the population lives on less than $2.50 (PPP) a day, down from 92.5% in 1981. This is much higher than the 80.5% in Sub-Saharan Africa.
- 75.6% of the population lives on less than $2 a day (PPP), which is around 20 rupees or $0.5 a day in nominal terms. It was down from 86.6%, but is still even more than the 73.0% in Sub-Saharan Africa.
- 24.3% of the population earned less than $1 (PPP, around $0.25 in nominal terms) a day in 2005, down from 42.1% in 1981.
- 41.6% of its population is living below the new international poverty line of $1.25 (PPP) per day, down from 59.8% in 1981. The World Bank further estimates that a third of the global poor now reside in India.

Today, more people can afford a bicycle than ever before. Some 40% of Indian households owns a bicycle, with ownership rates ranging from around 30% to 70% at state level. Housing is modest. According to Times of India, "a majority of Indians have per capita space equivalent to or less than a 10 feet x 10 feet room for their living, sleeping, cooking, washing and toilet needs." and "one in every three urban Indians lives in homes too cramped to exceed even the minimum requirements of a prison cell in the US." The average is 103 sq ft (9.6 m^2) per person in rural areas and 117 sq ft (10.9 m^2) per person in urban areas.

Around half of Indian children are malnourished. The proportion of underweight children is nearly double that of Sub-Saharan Africa. However, India has not had famines since the Green Revolution in the early 1970s. While poverty in India has reduced significantly, official figures estimate that 27.5% of Indians still lived below the national poverty line of $1 (PPP, around 10 rupees in nominal terms) a day in 2004-2005. A 2007 report by the state-run National Commission for Enterprises in the Unorganised Sector (NCEUS) found that 65% of Indians, or 750 million people, lived on less than 20 rupees per day with most working in "informal labour sector with no job or social security, living in abject poverty."

Since the early 1950s, successive governments have implemented various schemes, under planning, to alleviate poverty, that have met with partial success. All these programmes have relied upon the strategies of the *Food for work* programme and *National Rural Employment Programme* of the 1980s, which attempted to use the unemployed to generate productive assets and build rural infrastructure. In August 2005, the Indian parliament passed the *Rural Employment Guarantee Bill*, the largest programme of this type in terms of cost and coverage, which promises 100 days of minimum wage employment to every rural household in all the India's 600 districts. The question of whether economic reforms have reduced poverty or not has fuelled debates without generating any clear cut answers and has also put political pressure on further economic reforms, especially those involving the downsizing of labour and cutting agricultural subsidies.

Employment

Agricultural and allied sectors accounted for about 60% of the total workforce in 2003 same as in 1993–94. While agriculture has faced stagnation in growth, services have seen a steady growth. Of the total workforce, 8% is in the organised sector, two-thirds of which are in the public sector. The NSSO survey estimated that in 1999–2000, 106 million, nearly 10% of the population were unemployed and the overall unemployment rate was 7.3%, with rural areas doing marginally better (7.2%) than urban areas (7.7%). India's labour force is growing by 2.5% annually, but employment only at 2.3% a year.

Official unemployment exceeds 9%. Regulation and other obstacles have discouraged the emergence of formal businesses and jobs. Almost 30% of workers are casual workers who work only when they are able to get jobs and remain unpaid for the rest of the time. Only 10% of the workforce is in regular employment. India's labour regulations are heavy even by developing country standards and analysts have urged the government to abolish them.

Unemployment in India is characterized by chronic or disguised unemployment. Government schemes that target eradication of both poverty and unemployment (which in recent decades has sent millions of poor and unskilled people into urban areas in search of livelihoods) attempt to solve the problem, by providing financial assistance for setting up businesses, skill honing, setting up public sector enterprises, reservations in governments, etc. The decreased role of the public sector after liberalization has further underlined the need for focusing on better education and has also put political pressure on further reforms.

Child labour is a complex problem that is basically rooted in poverty. The Indian government is implementing the world's largest child labour elimination program, with primary education targeted for ~250 million. Numerous non-governmental and voluntary organizations are also involved. Special investigation cells have been set up in states to enforce existing laws banning employment of children (under 14) in hazardous industries.

The allocation of the Government of India for the eradication of child labour was $10 million in 1995-96 and $16 million in 1996-97. The allocation for 2007 is $21 million. In 2006, remittances from Indian migrants overseas made up $27 billion or about 3% of India's GDP.

Economic Trends

In the revised 2007 figures, based on increased and sustaining growth, more inflows into foreign direct investment, Goldman Sachs predicts that "from 2007 to 2020, India's GDP per capita in US$ terms will quadruple", and that the Indian economy will surpass the United States (in US$) by 2043.

In spite of the high growth rate, the report stated that India would continue to remain a low-income country for decades to come but could be a "motor for the world economy" if it fulfills its growth potential. Goldman Sachs has outlined 10 things that it needs to do in order to achieve its potential and grow 40 times by 2050.

These are;

1. improve governance
2. raise educational achievement
3. increase quality and quantity of universities
4. control inflation
5. introduce a credible fiscal policy
6. liberalize financial markets
7. increase trade with neighbours
8. increase agricultural productivity
9. improve infrastructure and
10. improve environmental quality.

Issues

Agriculture: Slow agricultural growth is a concern for policymakers as some two-thirds of India's people depend on rural employment for a living. Current agricultural practices are neither economically nor environmentally sustainable and India's yields for many agricultural commodities are low. Poorly maintained irrigation systems and almost universal lack of good extension services are among the factors responsible. Farmers' access to markets is hampered by poor roads, rudimentary market infrastructure, and excessive regulation.

– *World Bank: "India Country Overview 2008"*

The low productivity in India is a result of the following factors:

- According to "India: Priorities for Agriculture and Rural Development" by World Bank, India's large agricultural subsidies are hampering productivity-enhancing investment. Overregulation of agriculture has increased costs, price risks and uncertainty. Government interventions in labour, land, and credit markets are hurting the market. Infrastructure and services are inadequate.
- Illiteracy, slow progress in implementing land reforms and inadequate or inefficient finance and marketing services for farm produce.
- The average size of land holdings is very small (less than 20,000 m^2) and is subject to fragmentation, due to land ceiling acts and in some cases, family disputes. Such small holdings are often over-manned, resulting in disguised unemployment and low productivity of labour.
- Adoption of modern agricultural practices and use of technology is inadequate, hampered by ignorance of such practices, high costs and impracticality in the case of small land holdings.
- World Bank says that the allocation of water is inefficient, unsustainable and inequitable. The irrigation infrastructure is deteriorating. Irrigation facilities are inadequate, as revealed by the fact that only 52.6% of the land was irrigated in 2003–04, which result in farmers still being dependent on rainfall, specifically the Monsoon season. A good monsoon results in a robust growth for the economy as a whole, while a poor monsoon leads to a sluggish growth. Farm credit is regulated by NABARD, which is the statutory apex agent for rural development in the subcontinent.

India has many farm insurance companies that insure wheat, fruit, rice and rubber farmers in the event of natural disasters or catastrophic crop failure, under the supervision of the Ministry of Agriculture. One notable company that provides all of these insurance policies is Agriculture Insurance Company of India and it alone insures almost 20 million farmers.

India's population is growing faster than its ability to produce rice and wheat. The most important structural reform for self-sufficiency is the ITC Limited plan to connect 20,000 villages to the Internet by 2013. This will provide farmers with up to date crop prices for the first time, which should minimise losses incurred from neighbouring producers selling early and in turn facilitate investment in rural areas.

Corruption

Corruption has been one of the pervasive problems affecting India. The economic reforms of 1991 reduced the red tape, bureaucracy and the *Licence Raj* that had strangled private enterprise and was blamed by Chakravarthi Rajagopalachari for the corruption and inefficiencies. Yet, a 2005 study by Transparency International (TI) India found that more than half of those surveyed had firsthand experience of paying bribe or peddling influence to get a job done in a public office.

The Right to Information Act (2005) and equivalent acts in the Indian states, that require government officials to furnish information requested by citizens or face punitive action, computerisation of services and various central and state government acts that established vigilance commissions have considerably reduced corruption or at least have opened up avenues to redress grievances. The 2009 report by Transparency International ranks India at 84th place and states that significant improvements were made by India in reducing corruption.

Government

The current government has concluded that most spending fails to reach its intended recipients. Lant Pritchett calls India's public sector "one of the world's top ten biggest problems — of the order of AIDS and climate change". The Economist's 2008 article about the Indian civil service stated that the Indian central government employs around 3 million people, including "vast armies of paper-shuffling peons". At local level, administration can be worse. It is not unheard of that a majority of a state's assembly seats can be held by convicted criminals. One study found that 25% of public sector teachers and 40% of public sector medical workers could not be found at the workplace. India's absence rates are one of the worst in the world.

Education

India has made huge progress in terms of increasing primary education attendance rate and expanding literacy to approximately two thirds of the population. The right to education at elementary level has been made one of the fundamental rights under the Eighty-Sixth Amendment of 2002. However, the literacy rate of 65% is still lower than the worldwide average and the country suffers from a high dropout rate.

Infrastructure

In the past, development of infrastructure was completely in the hands of the public sector and was plagued by corruption, bureaucratic inefficiencies, urban-bias and an inability to scale investment. India's low spending on power, construction, transportation, telecommunications and real estate, at $31 billion or 6% of GDP in 2002 had prevented India from sustaining higher growth rates. This has prompted the government to partially open up infrastructure to the private sector allowing foreign investment which has helped in a sustained growth rate of close to 9% for the past six quarters.

Some 600 million Indians have no mains electricity at all. While 80% of Indian villages have at least an electricity line, just 44% of rural households have access to electricity. According to a sample of 97,882 households in 2002, electricity was the main source of lighting for 53% of rural households compared to 36% in 1993. Some half of the electricity is stolen, compared with 3% in China.

The stolen electricity amounts to 1.5% of GDP. Almost all of the electricity in India is produced by the public sector. Power outages are common. Many buy their own power generators to ensure electricity supply. As of 2005 the electricity production was at 661.6 billion kWh with oil production standing at 785,000 bbl/day. In 2007, electricity demand exceeded supply by 15%. Multi Commodity Exchange has tried to get a permit to offer electricity future markets.

Indian Road Network is developing. Trucking goods from Gurgaon to the port in Mumbai can take up to 10 days. India has the world's third largest road network. Container traffic is growing at 15% a year. Some 60% of India's container traffic is handled by the Jawaharlal Nehru Port Trust in Navi Mumbai.

Internet use is rare; there were only 7.57 million broadband lines in India in November 2009, however it is still growing at slower rate and is expected to boom after the launch of 3G and wimax services.

Most urban cities have good water supply water 24 hours a day, while some smaller cities face water shortages in summer season. A World Bank report says it is an institutional problem in water agencies, or "how the agency is embedded in the relationships between politics and the citizens who are the consumers."

Labour Laws

India's labour regulations — among the most restrictive and complex in the world — have constrained the growth of the formal manufacturing sector where these laws have their widest application. Better designed labour regulations can attract more labour-intensive investment and create jobs for India's unemployed millions and those trapped in poor quality jobs. Given the country's momentum of growth, the window of opportunity must not be lost for improving the job prospects for the 80 million new entrants who are expected to join the work force over the next decade.

– *World Bank: India Country Overview 2008.*

India's restrictive labour regulations hamper the large-scale creation of formal industrial jobs.

India ranked 133th on the Ease of Doing Business Index 2010, behind countries such as China (89th), Pakistan (85th), and Nigeria (125th). The Constitution provides protection of child labour, slavory, equality of opportunities and forced labour etc. in form of fundamental rights, but the implementation of provisions cited is a big question mark.

Economic Disparities

Lagging states need to bring more jobs to their people by creating an attractive investment destination. Reforming cumbersome regulatory procedures, improving rural connectivity, establishing law and order, creating a stable platform for natural resource investment that balances business interests with social concerns, and providing rural finance are important.

– *World Bank: India Country Overview 2008*

One of the critical problems facing India's economy is the sharp and growing regional variations among India's different states and territories in terms of per capita income, poverty, availability of infrastructure and socio-economic development. Six low-income states-Bihar, Chhattisgarh, Jharkhand, Madhya Pradesh, Orissa and Uttar Pradesh-are home to more than one third of India's population. Between 1999 and 2008, the annualized growth rates for Maharashtra (9%),

Gujarat (8.8%), Haryana (8.7%), or Delhi (7.4%) were much higher than for Bihar (5.1%), Uttar Pradesh (4.4%), or Madhya Pradesh (3.5%). However, In 2009-10, Bihar witnessed a growth of about 12.6%, and ended up becoming the 'next big economy in india', followed by Gujarat with a growth of 11.3%. Poverty rates in rural Orissa (43%) and rural Bihar (40%) are some of the worst in the world. On the other hand, rural Haryana (5.7%) and rural Punjab (2.4%) compare well with middle-income countries.

The five-year plans have attempted to reduce regional disparities by encouraging industrial development in the interior regions, but industries still tend to concentrate around urban areas and port cities After liberalization, the more advanced states are better placed to benefit from them, with infrastructure like well developed ports, urbanisation and an educated and skilled workforce which attract manufacturing and service sectors. The union and state governments of backward regions are trying to reduce the disparities by offering tax holidays, cheap land, etc., and focusing more on sectors like tourism, which although being geographically and historically determined, can become a source of growth and is faster to develop than other sectors.

Environment and Health

On Yale and Columbia's Environmental Performance Index, India's score is 21/100 on sanitation, compared with 67/100 for the region and 48/100 for the country income group.

Global Financial Crisis

Among the large economies, India has been one of the least affected by the 2008 global financial crisis and the recession that followed, with the stimulus measures implemented by the country relatively small. Economic growth rates did decline for several quarters, but not as much as in most other countries and a domestic recession was no more than a remote possibility. The relatively better structural balance of the Indian economy, dominated by domestic consumption, deflected the collateral damage from the collapse of the global financial markets and the steep fall in global trade volumes. More importantly, some of the programs initiated by the Indian government even before the crisis inadvertently proved to be the most effective stimulus measures.The most significant of the nationwide programs, which helped sustain domestic demand in India, was a federal initiative to supply temporary employment to the most economically distressed in rural areas. The program offered assured work to adult members in every poor rural family for a minimum of 100 days a year. Though it was stained by poor implementation and allegations of widespread corruption in its early days, the program has since been judged a moderate success, which is significant in a country where government initiatives have poor track record. In early 2008, the federal government also paid off the nearly $15 billion debt owed by more than 36 million small farmers to commercial banks. This initiative was designed to provide financial support to farmers who were worried by low farm produce prices, and constant droughts in several parts of the country. In the short-term, the program added substantially to the cash available to rural consumers, which supported domestic demand.

In 2009, India's nominal GDP stood at US$1.243 trillion, which makes it the eleventh-largest economy in the world. If PPP is taken into account, India's economy is the fourth largest in the world at US$3.561 trillion, corresponding to a per capita income of US$3,100. The country ranks 139th in nominal GDP per capita and 128th in GDP per capita at PPP. With an average annual GDP growth rate of 5.8% for the past two decades, India is one of the fastest growing economies in the world. India has the world's second largest labour force, with 516.3 million people. In terms of output, the agricultural sector accounts for 28% of GDP; the service and industrial sectors make up 54% and 18% respectively. Major agricultural products include rice, wheat, oilseed, cotton, jute, tea, sugarcane, potatoes; cattle, water buffalo, sheep, goats, poultry; fish. Major industries include textiles, telecommunications, chemicals, food processing, steel, transport equipment, cement, mining, petroleum, machinery, software. India's trade has reached a relatively moderate share of 24% of GDP in 2006, up from 6% in 1985. In 2008, India's share of world trade was about 1.68%. Major exports include petroleum products, textile goods, gems and jewelry, software, engineering goods, chemicals, and leather manufactures. Major imports include crude oil, machinery, gems, fertilizer, chemicals.

From the 1950s to the 1980s, India followed socialist-inspired policies. The economy was shackled by extensive regulation, protectionism, and public ownership, leading to pervasive corruption and slow growth. In 1991, the nation liberalised its economy and has since moved towards a market-based system. The policy change in 1991 came after an acute balance of payments crisis, and the emphasis since then has been to use foreign trade and foreign investment as integral parts of India's economy.

In the late 2000s, India's economic growth averaged 7.5% a year. Over the past decade, hourly wage rates in India have more than doubled. In 2009, the Global Competitiveness Report ranked India 16th in financial market sophistication, 24th in banking sector, 27th in business sophistication and 30th in innovation; ahead of several advanced economies. Seven of the world's top 15 technology outsourcing companies are based in India and the country is viewed as the second most favourable outsourcing destination after the United States.

Despite India's impressive economic growth over recent decades, it still contains the largest concentration of poor people in the world. The percentage of people living below the World Bank's international poverty line of $1.25 a day (PPP, in nominal terms Rs. 21.6 a day in urban areas and Rs. 14.3 in rural areas in 2005) decreased from 60% in 1981 to 42% in 2005. Since 1991, inter-state economic inequality in India has consistently grown; the per capita net state domestic product of India's richest states is about 3.2 times that of the poorest states. Even though India has avoided famines in recent decades, half of children are underweight and about 46% of Indian children under the age of three suffer from malnutrition.

A 2007 Goldman Sachs report projected that "from 2007 to 2020, India's GDP per capita will quadruple," and that the Indian GDP will surpass that of the United States before 2050, but India "will remain a low-income country for several decades, with per capita incomes well below its other BRIC peers." Although the Indian economy has grown steadily over the last two decades; its growth has been uneven when comparing different social groups, economic groups, geographic regions, and rural and urban areas. The World Bank suggests that India must continue to focus on public sector reform, infrastructure, agricultural and rural development, removal of labour regulations, improvement in transport, energy security, and health and nutrition.

Demographics

With an estimated population of 1.2 billion, India is the world's second most populous country. The last 50 years have seen a rapid increase in population due to medical advances and massive increase in agricultural productivity due to the "green revolution". India's urban population increased 11-fold during the twentieth century and is increasingly concentrated in large cities. By 2001 there were 35 million-plus population cities in India, with the largest cities, with a population of over 10 million each, being Mumbai, Delhi and Kolkata. However, as of 2001, more than 70% of India's population continues to reside in rural areas.

India is the world's most culturally, linguistically and genetically diverse geographical entity after the African continent. India is home to two major linguistic families: Indo-Aryan (spoken by about 74% of the population) and Dravidian (spoken by about 24%). Other languages spoken in India come from the Austro-Asiatic and Tibeto-Burman linguistic families. Neither the Constitution of India, nor any Indian law defines any *national language*. Hindi, with the largest number of speakers, is the official language of the union. English is used extensively in business and administration and has the status of a 'subsidiary official language;' it is also important in education, especially as a medium of higher education. In addition, every state and union territory has its own official languages, and the constitution also recognises in particular 21 "scheduled languages".

As per the 2001 census, over 800 million Indians (80.5%) were Hindu. Other religious groups include Muslims (13.4%), Christians (2.3%), Sikhs (1.9%), Buddhists (0.8%), Jains (0.4%), Jews, Zoroastrians and Bahais. Tribals constitute 8.1% of the population. India has the third-highest Muslim population in the world and has the highest population of Muslims for a non-Muslim majority country.

India's literacy rate is 64.8% (53.7% for females and 75.3% for males). The state of Kerala has the highest literacy rate at 91% while Bihar has the lowest at 47%. The national human sex ratio is 944 females per 1,000 males. India's median age is 24.9, and the population growth rate of 1.38% per annum; there are 22.01 births per 1,000 people per year. Though India has one of the world's most diverse and modern healthcare systems, the country continues to face several public health-related challenges. According to the World Health Organization, 900,000 Indians die each year from drinking contaminated water and breathing in polluted air. There are about 60 physicians per 100,000 people in India.

Culture

India's culture is marked by a high degree of syncretism and cultural pluralism. India's cultural tradition dates back to 8,000 BCE and has a continuously recorded history for over 2,500 years. With its roots based in the Indus Valley Tradition, the Indian culture took a distinctive shape during the 11th century BCE Vedic age which laid the foundation of Hindu philosophy, mythology, literary tradition and beliefs and practices, such as Dharma, Karma, Yoga and Mokca. It has managed to preserve established traditions while absorbing new customs, traditions, and ideas from invaders and immigrants and spreading its cultural influence to other parts of Asia, mainly South East and East Asia.

Indian religions form one of the most defining aspects of Indian culture. Major dharmic religions which were founded in India include Hinduism, Buddhism and Jainism. Considered to be a successor to the ancient Vedic religion, Hinduism has been shaped by several schools of thoughts such as the *Advaita Vedanta*, the *Yoga Sutras* and the *Bhakti* movement. Buddhism originated in India in 5th century BCE and prominent early Buddhist schools, such as *Theravada* and *Mahayana*, gained dominance during the Maurya Empire. Though Buddhism entered a period of gradual decline in India 5th century CE onwards, it played an influential role in shaping Indian philosophy and thought.

Traditional Indian society is defined by relatively strict social hierarchy. The Indian caste system describes the social stratification and social restrictions in the Indian subcontinent, in which social classes are defined by thousands of endogamous hereditary groups, often termed as *jatis* or castes. Several influential social reform movements, such as the *Bramho Shomaj*, the *Arya Samaja* and the *Ramakrishna* Mission, have played a pivotal role in the emancipation of Dalits (or "untouchables") and other lower-caste communities in India. However, the majority of Dalits continue to live in segregation and are often persecuted and discriminated against.

Traditional Indian family values are highly respected, and multi-generational patriarchal joint families have been the norm, although nuclear family are becoming common in urban areas. An overwhelming majority of Indians have their marriages arranged by their parents and other respected family members, with the consent of the bride and groom. Marriage is thought to be for life, and the divorce rate is extremely low. Child marriage is still a common practice, with half of women in India marrying before the legal age of 18.

Indian cuisine is characterised by a wide variety of regional styles and sophisticated use of herbs and spices. The staple foods in the region are rice (especially in the south and the east) and wheat (predominantly in the north). Spices, such as black pepper which are now consumed world wide, are originally native to the Indian subcontinent. Chili pepper, which was introduced by the Portuguese, is also widely used in Indian cuisine.

Traditional Indian dress varies across the regions in its colours and styles and depends on various factors, including climate. Popular styles of dress include draped garments such as sari for women and dhoti or lungi for men; in addition, stitched clothes such as salwar kameez for women and kurta-pyjama and European-style trousers and shirts for men, are also popular.

Many Indian festivals are religious in origin, although several are celebrated irrespective of caste and creed. Some popular festivals are Diwali, Ganesh Chaturthi, Ugadi, Thai Pongal, Holi, Onam, Vijayadashami, Durga Puja, Eid ul-Fitr, Bakr-Id, Christmas, Buddha Jayanti and Vaisakhi. India has three national holidays which are observed in all states and union territories — Republic Day, Independence Day and Gandhi Jayanthi. Other sets of holidays, varying between nine and twelve, are officially observed in individual states. Religious practices are an integral part of everyday life and are a very public affair. Indian architecture is one area that represents the diversity of Indian culture. Much of it, including notable monuments such as the Taj Mahal and other examples of Mughal architecture and South Indian architecture, comprises a blend of ancient and varied local traditions from several parts of the country and abroad. Vernacular architecture also displays notable regional variation.

Indian music covers a wide range of traditions and regional styles. Classical music largely encompasses the two genres – North Indian Hindustani, South Indian Carnatic traditions and their various offshoots in the form of regional folk music. Regionalised forms of popular music include filmi and folk music; the syncretic tradition of the *bauls* is a well-known form of the latter.

Indian dance too has diverse *folk* and *classical* forms. Among the well-known folk dances are the *bhangra* of the Punjab, the *bihu* of Assam, the *chhau* of West Bengal, Jharkhand, *Sambalpuri* of Orissa, the *ghoomar* of Rajasthan and the *Lawani* of Maharashtra. Eight dance forms, many with narrative forms and mythological elements, have been accorded classical dance status by India's *National Academy of Music, Dance, and Drama*. These are: *Bharatanatyam* of the state of Tamil Nadu, *Kathak* of Uttar Pradesh, *Kathakali* and *Mohiniyattam* of Kerala, *Kuchipudi* of Andhra Pradesh, *Manipuri* of Manipur, *Odissi* of Orissa and the *Sattriya* of Assam. Theatre in India often incorporates music, dance, and improvised or written dialogue. Often based on Hindu mythology, but also borrowing from medieval romances, and news of social and political events, Indian theatre includes the *Bhavai* of state of Gujarat, the *Jatra* of West Bengal, the *Nautanki* and *Ramlila* of North India, the *tamasha* of Maharashtra, the *Burrakatha* of Andhra Pradesh, the *Terukkuttu* of Tamil Nadu, and the *Yakshagana* of Karnataka. The Indian film industry is the largest in the world. Bollywood, based in Mumbai, makes commercial Hindi films and is the most prolific film industry in the world. Established traditions also exist in Assamese, Bengali, Kannada, Malayalam, Marathi, Tamil, and Telugu language cinemas.

The earliest works of Indian literature were transmitted orally and only later written down. These included works of Sanskrit literature – such as the early Vedas, the epics Mahabharata and Ramayana, the drama *Abhijnanauakuntalam* (The Recognition of Uakuntala), and poetry such as the *Mahakavya* – and the Tamil language *Sangam* literature. Among Indian writers of the modern era active in Indian languages or English, Rabindranath Tagore won the Nobel Prize in 1913. Southeast Asia (or Southeastern Asia) is a subregion of Asia, consisting of the countries that are geographically south of China, east of India and north of Australia. The region lies on the intersection of geological plates, with heavy seismic and volcanic activity. Southeast Asia consists of two geographic regions: the Asian mainland, and island arcs and archipelagoes to the east and southeast. The mainland section consists of Burma (Myanmar), Cambodia, Laos, Thailand, Vietnam and Peninsular Malaysia while the maritime section consists of Brunei, East Malaysia, East Timor, Indonesia, the Philippines, and Singapore.

Austronesian peoples predominate in this region. The major religions are Islam and Buddhism, followed by Christianity. However a wide variety of religions are found throughout the region, including many Hindu and animist-influenced practices.

Divisions

Political

Definitions of "Southeast Asia" vary, but most definitions include the area represented by the countries:

- Brunei
- Burma (Myanmar)
- Cambodia
- East Timor (Timor-Leste)
- Indonesia
- Laos
- Malaysia
- Philippines
- Singapore
- Thailand
- Vietnam.

All of the above excluding Timor-Leste are members of the Association of Southeast Asian Nations (commonly abbreviated

ASEAN.) The area, together with part of South Asia, was widely known as the East Indies or simply the Indies until the twentieth century. Christmas Island and the Cocos (Keeling) Islands are considered part of Southeast Asia though they are governed by Australia. Sovereignty issues exist over some islands in the South China Sea. Papua is politically part of Southeast Asia through Indonesia, although geographically it is often considered as part of Oceania. As of 2009, Papua New Guinea has stated that it might join ASEAN, indicating a possible switch in its geographic locale.

Geographical

The eastern parts of Indonesia and East Timor (east of Wallace Line) are considered to be geographically parts of Oceania.

Southeast Asia is geographically divided into two subregions, namely Mainland Southeast Asia (or Indochina) and Maritime Southeast Asia (or the similarly defined Malay Archipelago) (Indonesian: *Nusantara*).

Mainland Southeast Asia includes:

- Myanmar (Burma)
- Cambodia
- Laos
- Thailand
- Vietnam
- Peninsular Malaysia.

Maritime Southeast Asia includes:

- East Malaysia
- Brunei
- Indonesia
- Philippines
- Singapore
- Timor-Leste.

The Andaman and Nicobar Islands of India are geographically considered part of Southeast Asia. Bangladesh and the Seven Sister States of India are culturally part of Southeast Asia and sometimes considered both South Asian and Southeast Asian. The Seven Sister States of India are also geographically part of southeast Asia. Hainan Island and several other southern Chinese regions such as Yunnan, Guizhou and Guangxi are considered both East Asian and Southeast

Asian. The rest of New Guinea is sometimes included so are Palau, Guam, and the Northern Mariana Islands, which were all part of the Spanish East Indies.

Countries and Territories Data

History

Homo sapiens reached the region by around 45,000 years ago. *Homo floresiensis* seems to have shared some islands with modern humans until only 12,000 years ago, when they became extinct. Austronesian people, who form the majority of the modern population in Indonesia and the Philippines, migrated to South East Asia from Taiwan. They arrived in Indonesia around 2000 BCE, and as they spread through the archipelago, confined the native Melanesian peoples to the far eastern regions.

Solheim and others have shown evidence for a *Nusantao* (*Nusantara*) maritime trading network ranging from Vietnam to the rest of the archipelago as early as 5000 BCE to 1 CE. The peoples of Southeast Asia, especially those of Austronesian descent, have been seafarers for thousands of years, some reaching the island of Madagascar. Their vessels, such as the vinta, were ocean-worthy. Magellan's voyage records how much more maneuvreable their vessels were, as compared to the European ships.

Passage through the Indian Ocean aided the colonization of Madagascar by the Austronesian people, as well as commerce between West Asia and Southeast Asia. Gold from Sumatra is thought to have reached as far west as Rome, while slaves from the Sulu Sea was believed to have been used in Magellan's voyage as a translator.

Originally most people were animist. This was later replaced by Brahmanic Hinduism. Theravada Buddhism soon followed in 525. In 1400s, Islamic influences began to enter. This forced the last Hindu court in Indonesia to retreat to Bali. In Mainland Southeast Asia, Myanmar, Cambodia and Thailand retained the Theravada form of Buddhism, brought to them from Sri Lanka. This type of Buddhism was fused with the Hindu-influenced Khmer culture.

Indianized Kingdoms

Very little is known about Southeast Asian religious beliefs and practices before the advent of Indian merchants and religious influences from the second century BCE onwards. Prior to the 13th century, Hinduism and Buddhism were the main religions in Southeast Asia.

The Jawa Dwipa Hindu kingdom in Java and Sumatra existed around 200 BCE. The history of the Malay-speaking world begins with the advent of Indian influence, which dates back to at least the 3rd century BC. Indian traders came to the archipelago both for its abundant forest and maritime products and to trade with merchants from China, who also discovered the Malay world at an early date. Both Hinduism and Buddhism were well established in the Malay Peninsula by the beginning of the 1st century CE, and from there spread across the archipelago.

Cambodia was first influenced by Hinduism during the beginning of the Funan kingdom. Hinduism was one of the Khmer Empire's official religions. Cambodia is the home to one of the only two temples dedicated to Brahma in the world. Angkor Wat is also a famous Hindu temple of Cambodia. The Champa civilization was located in what is today central Vietnam, and was a highly indianized Hindu Kingdom.

The Majapahit Empire was an Indianized kingdom based in eastern Java from 1293 to around 1500. Its greatest ruler was Hayam Wuruk, whose reign from 1350 to 1389 marked the empire's peak when it dominated other kingdoms in the southern Malay Peninsula, Borneo, Sumatra, and Bali. Various sources such as the Nagarakertagama also mention that its influence spanned over parts of Celebes, the Moluccas islands, and some areas of western Papua, making it the largest empire to ever exist in Southeast Asian history.

The Cholas excelled in maritime activity in both military and the mercantile fields. Their raids of Kedah and the Srivijaya, and their continued commercial contacts with the Chinese Empire, enabled them to influence the local cultures. Many of the surviving examples of the Hindu cultural influence found today throughout the Southeast Asia are the result of the Chola expeditions.

Islamization of Southeast Asia

In the 11th century, a turbulent period occurred in the history of Maritime Southeast Asia, the Indian Chola navy crossed the ocean and attacked the Srivijaya kingdom of Sangrama Vijayatungavarman in Kadaram (Kedah), the capital of the powerful maritime kingdom was sacked and the king was taken captive. Along with Kadaram, Pannai in present day Sumatra and Malaiyur and the Malayan peninsula were attacked too. Soon after that, the king of Kedah Phra Ong Mahawangsa became the first ruler to abandon the traditional Hindu faith, and converted to Islam with the Sultanate of Kedah established in year 1136. Samudera Pasai converted to Islam in the year 1267, the King of Malacca Parameswara married with princess of Pasai, the son became the first sultan of Malacca, soon Malacca became the centre of Islam study and maritime trade, other rulers followed suit. Indonesian religious leader and Islamic scholar Hamka (1908–1981) wrote in 1961:

> *"The development of Islam in Indonesia and Malaya is intimately related to a Chinese Muslim, Admiral Zheng He."*

There are several theories to the Islamization process in Southeast Asia. The first theory is trade. The expansion of trade among West Asia, India and Southeast Asia helped the spread of the religion as Muslim traders brought Islam to the region. The second theory is the role of missionaries or Sufis.

The Sufi missionaries played a significant role in spreading the faith by syncretising Islamic ideas with existing local beliefs and religious notions. Finally, the ruling classes embraced Islam and that further aided the permeation of the religion throughout the region. The ruler of the region's most important port, Malacca Sultanate, embraced Islam in the 15th century, heralding a period of accelerated conversion of Islam throughout the region as the religion provided a unifying force among the ruling and trading classes.

Trade and Colonization

China

Chinese merchants have traded with the region for a long time as evidence of Magellan's voyage records that Brunei possessed more cannon than the European ships so it appears that the Chinese fortified them. Malaysian legend has it that a Chinese Ming emperor sent a princess, Han Li Po, to Malacca, with a retinue of 500, to marry Sultan Mansur Shah after the emperor was impressed by the wisdom of the sultan. Han Li Po's well (constructed 1459) is now a tourist attraction there, as is Bukit Cina, where her retinue settled.

The strategic value of the Strait of Malacca, which was controlled by Sultanate of Malacca in the 15th and early 16th century, did not go unnoticed by Portuguese writer Duarte Barbosa, who in 1500 wrote "He who is lord of Malacca has his hand on the throat of Venice".

Europe

Western influence started to enter in the 1500s, with the arrival of the Portuguese and Spanish in Moluccas and the Philippines. Later the Dutch established the Dutch East Indies; the French Indochina; and the British Strait Settlements. All southeast Asian countries were colonized except for Thailand.

European explorers were reaching Southeast Asia from the west and from the east. Regular trade between the ships sailing east from the Indian Ocean and south from mainland Asia provided goods in return for natural products, such as honey and hornbill beaks from the islands of the archipelago.

Europeans brought Christianity allowing Christian missionaries to become widespread. Thailand also allowed Western science and technology to enter its country.

Japan

During World War II, the Imperial Japan invaded most of the former western colonies. The Showa occupation regime committed violent actions against indigenous civilians such as the Manila Massacre and the implementation of a system of forced labour, such as the one involving 4 to 10 million *romusha* in Indonesia. A later UN report stated that four million people died in Indonesia as a result of famine and forced labour during the Japanese occupation.

Present

Most countries in the region enjoy national autonomy. Democratic forms of government and the recognition of human rights are taking root. ASEAN provides a framework for the integration of commerce.

Conflicting territorial and maritime claims continue to exist, including the conflicting claims by Taiwan, China, and the Philippines over the Spratly Islands.

Geography

Geologically, the Indonesian archipelago is one of the most active vulcanological regions in the world. Geological uplifts in the region have also produced some impressive mountains, culminating in Puncak Jaya in Papua, Indonesia at 5,030 metres (16,024 ft), on the island of New Guinea, it is the only place where ice glacier can be found in Southeast Asia.

While the second tallest peak is Mount Kinabalu in Sabah, Malaysia on the island of Borneo with a height of 4,101 meters (13,455 ft). The tallest mountain in the Southeast Asia is Hkakabo Razi at 5,967 meters and can be found in northern Myanmar. The largest archipelago in the world by size is Indonesia.

Boundaries

The Australian continent defines a region adjacent to Southeast Asia, which is also politically separated from the countries of Southeast Asia. But a cultural touch point lies between Papua New Guinea and the Indonesian region of Papua and West Papua, which shares the island of New Guinea with Papua New Guinea.

Climate

The climate in Southeast Asia is mainly tropical–hot and humid all year round with plentiful rainfall. Southeast Asia has a wet and dry season caused by seasonal shift in winds or monsoon.

The tropical rain belt causes additional rainfall during the monsoon season. The rain forest is the second largest on earth (with the Amazon being the largest). An exception to this type of climate and vegetation is the mountain areas in the northern region, where high altitudes lead to milder temperatures and drier landscape. Other parts fall out of this climate because they are desert like.

Environment

All of Southeast Asia falls within the warm, humid tropics, and its climate generally can be characterized as monsoonal. The animals of Southeast Asia are diverse; on the islands of Borneo and Sumatra, the Orangutan (man of the forest), the Asian Elephant, the Malayan tapir, the Sumatran Rhinoceros and the Bornean Clouded Leopard can be also found. Six subspecies of the Binturong or *bearcat* exist in the region, though the one endemic to the island of Palawan is now classed as vulnerable.

The Komodo Dragon is the largest living species of lizard and inhabits the islands of Komodo, Rinca, Flores, and Gili Motang in Indonesia.

The Wild Asian Water Buffalo, and on various islands related dwarf species of Bubalus such as Anoa were once widespread in Southeast Asia, nowadays the Domestic Asian Water buffalo is common across the region, but its remaining relatives are rare and endangered.

The mouse deer, a small tusked deer as large as a toy dog or cat, can be found on Sumatra, Borneo and Palawan Islands. The gaur, a gigantic wild ox larger than even wild Water buffalo, is found mainly in Indochina.

Birds such as the peafowl and drongo live in this subregion as far east as Indonesia. The babirusa, a four-tusked pig, can be found in

Indonesia as well. The hornbill was prized for its beak and used in trade with China. The horn of the rhinoceros, not part of its skull, was prized in China as well.

The Indonesian Archipelago is split by the Wallace Line. This line runs along what is now known to be a tectonic plate boundary, and separates Asian (Western) species from Australasian (Eastern) species. The islands between Java/Borneo and Papua form a mixed zone, where both types occur, known as Wallacea. As the pace of development accelerates and populations continue to expand in Southeast Asia, concern has increased regarding the impact of human activity on the region's environment.

A significant portion of Southeast Asia, however, has not changed greatly and remains an unaltered home to wildlife. The nations of the region, with only few exceptions, have become aware of the need to maintain forest cover not only to prevent soil erosion but to preserve the diversity of flora and fauna. Indonesia, for example, has created an extensive system of national parks and preserves for this purpose. Even so, such species as the Javan rhinoceros face extinction, with only a handful of the animals remaining in western Java.

The shallow waters of the Southeast Asian coral reefs have the highest levels of biodiversity for the world's marine ecosystems, where coral, fish and molluscs abound. According to Conservation International, marine surveys suggest that the marine life diversity in the Raja Ampat area is the highest recorded on Earth. Diversity is considerably greater than any other area sampled in the Coral Triangle composed of Indonesia, Philippines and Papua New Guinea. The Coral Triangle is the heart of the world's coral reef biodiversity, making Raja Ampat quite possibly the richest coral reef ecosystems in the world. The whale shark, the world's largest species of fish and 6 species of pawikans can also be found in the South China Sea and the Pacific Ocean territories of the Philippines.

The trees and other plants of the region are tropical; in some countries where the mountains are tall enough, temperate-climate vegetation can be found. These rainforest areas are currently being logged-over, especially in Borneo.

While Southeast Asia is rich in flora and fauna, Southeast Asia is facing severe deforestation which causes habitat loss for various endangered species such as orangutan and the Sumatran tiger. Predictions have been made that more than 40% of the animal and plant species in Southeast Asia could be wiped out in the 21st century.

At the same time, haze has been a regular occurrence. The two worst regional hazes were in 1997 and 2006 in which multiple countries were covered with thick haze, mostly caused by "slash and burn" activities in Indonesia. In reaction, several countries in Southeast Asia signed the ASEAN Agreement on Transboundary Haze Pollution in order to combat haze pollution.

Economy

Even prior to the penetration of European interests, Southeast Asia was a critical part of the world trading system. The Ryukyu Kingdom often participated in maritime trade in Southeast Asia. A wide range of commodities originated in the region, but especially important were such spices as pepper, ginger, cloves, and nutmeg. The spice trade initially was developed by Indian and Arab merchants, but it also brought Europeans to the region. First Spaniards (Manila galleon) and Portuguese, then the Dutch, and finally the British and French became involved in this enterprise in various countries. The penetration of European commercial interests gradually evolved into annexation of territories, as traders lobbied for an extension of control to protect and expand their activities. As a result, the Dutch moved into Indonesia, the British into Malaya, the French into Indochina and the Spanish into the Philippines

While the region's economy greatly depends on agriculture, manufacturing and services are becoming more important. An emerging market, Indonesia is the largest economy in this region. Newly industrialized countries include the Philippines, Malaysia, and Thailand, while Singapore and Brunei are affluent developed economies. The rest of Southeast Asia is still heavily dependent on agriculture, but Vietnam is notably making steady progress in developing its industrial sectors. The region notably manufactures textiles, electronic high-tech goods such as microprocessors and heavy industrial products such as automobiles. Reserves of oil are also present in the region.

Seventeen telecommunications companies have contracted to build a new submarine cable to connect Southeast Asia to the U.S. This is to avoid disruption of the kind recently caused by the cutting of the undersea cable from Taiwan to the U.S. in a recent earthquake.

Tourism has been a key factor in economic development for many Southeast Asian countries, especially Cambodia. According to UNESCO, "tourism, if correctly conceived, can be a tremendous development tool and an effective means of preserving the cultural diversity of our planet." Since the early 1990s, "even the non-ASEAN nations such as Cambodia, Laos, Vietnam and Myanmar, where the income derived from tourism is low, are attempting to expand their own tourism industries." In 1995, Singapore was the regional leader in tourism receipts relative to GDP at over 8%. By 1998, those receipts had dropped to less than 6% of GDP while Thailand and Lao PDR increased receipts to over 7%. Since 2000, Cambodia has surpassed all other ASEAN countries and generated almost 15% of its GDP from tourism in 2006.

Demographics

Southeast Asia has an area of approximately 4,000,000 km (1.6 million square miles). As of 2004, more than 593 million people lived in the region, more than a fifth of them (125 million) on the Indonesian island of Java, the most densely populated large island in the world. Indonesia is the most populous country with 230 million people and also 4th most populous country in the world. The distribution of the religions and people is diverse in Southeast Asia and varies by country. Some 30 million overseas Chinese also live in Southeast Asia, most prominently in Christmas Island, Malaysia, the Philippines, Singapore, Indonesia and Thailand, and also, as the Hoa, in Vietnam.

Ethnic Groups

According to a recent Stanford genetic study, the Southeast Asian population is far from being homogeneous. Although primarily descendants of Austronesian, Tai, and Mon-Khmer-speaking immigrants who migrated from Southern China during the Bronze Age and Iron Age, there are overlays of Arab, Chinese, Indian, Polynesian and Melanesian genes.

There are also large pockets of intermarriage between indigenous Southeast Asians and those of Chinese descent. They form a substantial part of everyday life in countries such as Vietnam, Singapore, Thailand and the Philippines. Indonesia and Malaysia also has a few mixed Southeast Asian-Chinese populations. On the mainland the Khmer peoples of Cambodia remain as ancestors of earlier Pareoean peoples. Similarly, remnants of the Mon group are found in parts of Myanmar and Thailand; the ethnic mixture there has been produced by overlaying Tibeto-Burman and Tai, Lao, and Shan peoples. The contemporary Vietnamese population originated from the Red River area in the north and may be a mixture of Tai and Malay peoples. Added to these major ethnic groups are such less numerous peoples as the Karens, Chins, and Nagas in Myanmar, who have affinities with other Asiatic peoples. Insular Southeast Asia contains a mixture of descendants of Proto-Malay (Nesiot) and Pareoean peoples who were influenced by Malayo-Polynesian and other groups. In addition, Arabic, Indian, and Chinese influences have affected the ethnic pattern of the islands. In modern times, the Javanese are the largest ethnic group in Southeast Asia, with more than 86 million people, mostly concentrated in Java, Indonesia. In Myanmar, the Burmese account for more than two-thirds of the ethnic stock in this country, while ethnic Thais and Vietnamese account for about four-fifths of the respective populations of those countries. Indonesia is clearly dominated by the Javanese and Sundanese ethnic groups, while Malaysia is more evenly split between the Malays and the Chinese. Within the Philippines, the Tagalog, Cebuano, Ilocano, and Bicol groups are significant.

Religions

Islam is the most widely practiced religion in Southeast Asia, numbering approximately 240 million adherents which translate to about 40% of the entire population, with majorities in Brunei, Indonesia and Malaysia. Countries in Southeast Asia practice many different religions. *Mainland* Southeast Asian countries, which are, Thailand, Cambodia, Laos, Myanmar, Vietnam and Singapore practice predominantly Buddhism. Ancestor worship and Confucianism is also widely practiced in Vietnam and Singapore. In Maritime Southeast Asia, people living in Malaysia, western Indonesia and Brunei practice mainly Islam. Christianity is predominant in the Philippines because of Spanish colonization for more than 300 years, eastern Indonesia and East Timor. The Philippines has the largest Roman Catholic population followed very distantly by Vietnam. East Timor is also predominantly Roman Catholic due to a history of Portuguese rule.

The religious composition for each country is as follows. Some values are taken from the *CIA World Factbook*: Religions and peoples are diverse in Southeast Asia and not one country is homogeneous. In the world's most populous Muslim nation, Indonesia, Hinduism is dominant on islands such as Bali. Christianity also predominates in Philippines, New Guinea and Timor.

The phoenix who is the mount (vahanam) of Vishnu, is a national symbol in both Thailand and Indonesia; in the Philippines, gold images of Garuda have been found on Palawan; gold images of other Hindu gods and goddesses have also been found on Mindanao. Balinese Hinduism is somewhat different from Hinduism practiced elsewhere, as Animism and local culture is incorporated into it. Cambodia still involves today in practice Hindism for mixture toward the excessity Buddhism disciple of the numerous special occasion, wornship and belief with the root of previous powerful Khmer Empire. Christians can also be found throughout Southeast Asia; they are in the majority in East

Timor and the Philippines, Asia's largest Christian nation. In addition, there are also older tribal religious practices in remote areas of Sarawak in East Malaysia and Papua in eastern Indonesia. In Myanmar, Sakka (Indra) is revered as a *nat*. In Vietnam, Mahayana Buddhism is practiced, which is influenced by native animism but with strong emphasis on Ancestor Worship.

Languages

Each of the languages have been influenced by cultural pressures due to trade and historical colonization as well. Thus, for example, a Filipino, educated in English and Filipino, as well as in his native tongue (e.g., Visayan), might well speak another language, such as Spanish for historical reasons, or Chinese, Korean or Japanese for economic reasons; a Malaysian might well speak English, Chinese, Tamil as well as Malay as a second language.

The language composition for each country is as follows: (official languages are in bold.)

Andaman and Nicobar Islands	Nicobarese, Bengali, English, Hindi, Malayalam, Punjabi, Tamil, Telugu, Shompen, Andamanese languages, others
Brunei	Malay, English, Chinese, indigenous Borneian dialects Cambodia Khmer, English, French, Vietnamese, Thai, Chamic dialects, Chinese languages, others Christmas Island English, Chinese, Malay Cocos (Keeling) Islands English, Cocos Malay East Timor Tetum, Portuguese, Indonesian, English, Mambae, Makasae, Tukudede, Bunak, Galoli, Kemak, Fataluku, Baikeno, others Indonesia Indonesian, Acehnese, Batak, Minang, Sundanese, Javanese, Banjarese, Sasak, Tetum, Dayak, Minahasa, Toraja, Buginese, Halmahera, Ambonese, Ceramese; Dutch, Papuan languages, Chinese, others Laos Lao, Thai, Vietnamese, Hmong, Miao, Mien, Dao, Shan; French, English others Malaysia Malay,English, Mandarin, Chinese dialects, Indian languages, Sarawakian and Sabahan languages, others Myanmar (Burma) Burmese, Shan, Karen, Rakhine, Kachin, Chin, Mon, Chinese languages, Indian languages, others
Philippines	Filipino, English, Tagalog, Cebuano, Ilocano, Hiligaynon/ Ilonggo, Kapampangan, Bicol, Waray, Pangasinan, Spanish and Arabic (optional) Singapore English, Mandarin, Malay, Tamil, other Chinese languages, other Indian languages, Arabic dialects, others

Contd...

South China Sea Islands	English, Filipino, Malay, Mandarin Chinese, Vietnamese
Thailand	Thai, English, Chinese languages, Malay, Lao, Khmer, Isaan, Shan, Lue, Phutai, Mon, Mein, Hmong, Karen, Burmese, others Vietnam Vietnamese, English, Chinese languages, French, Khmer, mountain area languages (Mon-Khmer and Malayo-Polynesian, hmong)

Culture

Rice paddy agriculture has existed in Southeast Asia for thousands of years, ranging across the subregion. Some dramatic examples of these rice paddies populate the Banaue Rice Terraces in the mountains of Luzon in the Philippines. Maintenance of these paddies is very labour-intensive. The rice paddies are well-suited to the monsoon climate of the region.

Stilt houses can be found all over Southeast Asia, from Thailand and Laos, to Borneo, to Luzon in the Philippines, to Papua New Guinea.

The region has diverse metalworking, especially in Indonesia. This include weaponry, such as the distinctive kris, and musical instruments, such as the gamelan.

Influences

The region's chief cultural influences have been from either China or India or both, with Vietnam considered by far the most Chinese-influenced. Myanmar can be said to be influenced equally by both India and China. Western cultural influence is most pronounced in the Philippines, derived particularly from the period of Spanish rule.

As a rule, the peoples who ate with their fingers were more likely influenced by the culture of India, for example, than the culture of China, where the peoples first ate with chopsticks; tea, as a beverage, can be found across the region. The fish sauces distinctive to the region tend to vary.

The Arts

The arts of Southeast Asia have no affinity with the arts of other areas. Dance in much of Southeast Asia also includes movement of the hands, as well as the feet to express the emotion and meaning of dance upon the story that the ballerina going to tell the audience. Most of Southeast Asian confirmed the Dance into their court, according to Cambodian royal ballet represent them in earlier of 7th century before Khmer Empire which highly influenced by Indian Hinduism. Apsara

Dance, famous for its strongly hand and feet movement, is a great example of Hindism symbol dance. Puppetry and shadow plays were also a favoured form of entertainment in past centuries as the famous one known as Wayang from Indonesia. The Arts and Literature in some of Southeast Asia is quite influenced by Hinduism brought to them centuries ago.

The Tai, coming late into Southeast Asia, brought with them some Chinese artistic traditions, but they soon shed them in favour of the Khmer and Mon traditions, and the only indications of their earlier contact with Chinese arts were in the style of their temples, especially the tapering roof, and in their lacquerware.

In Indonesia, despite conversion to Islam opposed to certain forms of art, they retained many forms of Hindu influenced practices, cultures, arts and literatures. An example will be the Wayang Kulit (Shadow Puppet) and literatures like the Ramayana. This is also true for mainland Southeast Asia (excluding Vietnam). Dance movements, Hindu gods, arts were also fused into Thai, Khmer, Lao and Burmese cultures. It has been pointed out that Khmer and Indonesian classical arts were concerned with depicting the life of the gods, but to the Southeast Asian mind the life of the gods was the life of the peoples themselves—joyous, earthy, yet divine.

In Vietnam, the Vietnamese share many cultural similarities with the Chinese.

Music

Traditional music in Southeast Asia is as varied as its many ethnic and cultural divisions. Main styles of traditional music can be seen: Court music, folk music, music styles of smaller ethnic groups, and music influenced by genres outside the geographic region.

Of the court and folk genres, Gong-chime ensembles and orchestras make up the majority (the exception being lowland areas of Vietnam). *Gamelan* orchestras from Indonesia, *Piphat/Pinpeat* ensembles of Thailand & Cambodia and the *Kulintang* ensembles of the southern Philippines, Borneo, Sulawesi and Timor are the three main distinct styles of musical genres that have influenced other traditional musical styles in the region. String instruments also are popular in the region.

Writing

The history of Southeast Asia has led to a wealth of different authors, from both within and without writing about the region.

Originally, Indians were the ones who taught the native inhabitants about writing. This is shown through Brahmic forms of writing present in the region such as the Balinese script shown on split palm leaf called *lontar*, right:

> *The antiquity of this form of writing extends before the invention of paper around the year 100 in China. Note each palm leaf section was only several lines, written longitudinally across the leaf, and bound by twine to the other sections. The outer portion was decorated.*
>
> *The alphabets of Southeast Asia tended to be abugidas, until the arrival of the Europeans, who used words that also ended in consonants, not just vowels. Other forms of official documents, which did not use paper, included Javanese copperplate scrolls.*
>
> *This would have been more durable in the tropical climate of Southeast Asia.*

Bibliography

Belden C.: *Landscapes of the Sacred: Geography and Narrative in American Spirituality*, New York: Paulist Press, 1988.

Berry, J.K. : *Beyond Mapping, Concepts, Algorithms and Issues in GIS*, Fort Collins, CO: GIS World Books, 1993.

Bolstad, P. : *GIS Fundamentals, A first text on Geographic Information Systems, Second Edition*, White Bear Lake, MN, Eider Press, 2005.

Cronon, William: *Changes in the Land: Indians, Colonists, and the Ecology of New England*. New York: Hill and Wang, 1983.

Davis, Mike: *Ecology of Fear: Los Angeles and the Imagination of Disaster*. New York: Vintage, 1999.

Elangovan K. : *GIS, Fundamentals, Applications and Population Geography*, New India Publishing Agency, New Delhi 2006.

Harvey, David: *Explanation in Geography*. London: Edward Arnold, 1969.

Harvey, Francis : *A Primer of GIS, Fundamental Geographic and Cartographic Concepts,* The Guilford Press, 2008.

Jackson, Peter: *Maps of Meaning: An Introduction to Cultural Geography*. London and Boston: Unwin Hyman, 1989.

James S. Duncan: *Writing Worlds: Discourse, Text, and Metaphor in the Representation of Landscape*. London and New York: Routledge, 1992.

James, T.: *The Best Poor Man's Country: A Geographical Study of Early Southeastern Pennsylvania*. Baltimore: Johns Hopkins Press, 1972.

King, Anthony D.: *Buildings and Society: Essays on the Social Development of the Built Enviroment*. London: Routledge & Kegan Paul, 1980.

Lewis, Peirce. *New Orleans: The Making of an Urban Landscape*. Cambridge, MA: Ballinger Publishing Co., 1976. Annotation

Liam Kennedy: *Urban Space and Representation*. Sterling, VA: Pluto Press, 2000.

Lung, Y., and van Tulder, R. : *Cars, Carriers of Regionalism*, Palgrave Macmillan, New York, 2004.

McDonnell, R.A. : *Principles of Geographical Information Systems*, Oxford University Press, Oxford, 1998.

Michael P.: *The Making of the American Landscape*. Boston: Unwin Hyman, 1990.

Michael, H.: *City Form and Natural Process: Towards a New Urban Vernacular*. London: Croom Helm, 1984.

Peter, G.: *The Slow Plague: A Geography of the AIDS Pandemic*. Oxford and Cambridge, MA: Blackwell Publishers, 1993.

Ray Allen: *The Far Western Frontier, 1830-1860*. The New American Nation Series. New York: Harper, 1956.

Richard SCASE : *Managing Creativity,* Open University Press, 2000.

Stephen, D.: *Fields of Vision: Landscape Imagery and National Identity in England and the United States*. Cambridge: Polity, 1993.

Tulder, R. : *Cars, Carriers of Regionalism*, Palgrave Macmillan, New York, 2004.

Victor, A. : *Geography of Art and Culture*, Contributions to Economic Analysis, Elsevier, 2004.

Watkins, Eric: *The Middle Eastern Environment*, London, The British Society for Middle Eastern Studie, 1995.

William G. BOWEN : *Performing Arts, The Economic Dilemma,* Thousand Oaks, SAGE, London, 2002.

Worboys, Michael, and Matt Duckham. : *GIS, a Computing Perspective*, Boca Raton, CRC Press, 2004.

Index

❑❑❑